250 Years of Progress

Fire and Emergency

Halifax Regional

Allison Lawlor

Copyright © Halifax Regional Fire and Emergency Service, 2005

All rights reserved. No part of this book may be reproduced, stored in a retrieval system or transmitted in any form or by any means without the prior written permission from the publisher, or, in the case of photocopying or other reprographic copying, permission from Access Copyright, 1 Yonge Street, Suite 1900, Toronto, Ontario M5E 1E5.

Nimbus Publishing Limited
PO Box 9166
Halifax, NS B3K 5M8
(902) 455-4286

Printed and bound in Canada

Design: Kathy Kaulbach, Paragon Design Group

Library and Archives Canada Cataloguing in Publication

    Lawlor, Allison, 1971-
    250 years of progress : Halifax regional fire and emergency services / Allison Lawlor
    Two hundred fifty years of progress

Includes bibliographical references.
ISBN 1-55109-541-6

TH9507.H28L38 2005   C813/.6 22   C2005-905893-5

We acknowledge the financial support of the Government of Canada through the Book Publishing Industry Development Program (BPIDP) and the Canada Council, and of the Province of Nova Scotia through the Department of Tourism, Culture and Heritage for our publishing activities.

Acknowledgements

I WOULD FIRST LIKE TO EXTEND SPECIAL THANKS to Dan Soucoup, Sandra McIntyre, and Heather Bryan at Nimbus Publishing, as well as Jeanette MacKay, my first link to the Halifax Regional Fire and Emergency, who for months helped me gather information and lead me to the right people. Thanks are also due to Darlene Ellis, who seems to know everyone and almost everything about the fire service and Captain Joe Ryan's crew at Station 2 on University Avenue, who took me out for my first ride in a fire engine and let me experience what it was like to spend twenty-four hours in their shoes.

My gratitude also goes to Mike Parker and the exhaustive research he put into his book The Smoke-Eaters: A History of Firefighting in Nova Scotia; Robert Walsh, who introduced me to the history of Dartmouth's fire department and provided me with great resources; and the Halifax Public Libraries and the Nova Scotia Archives and Records Management for their wonderful collections of books, newspapers, and photographs.

I especially thank Don Snider who was incredibly generous with his time, knowledge, and resources—but most of all, for sharing his love of firefighting history and the admirable work he has done to preserve it.

My deep appreciation goes to all the volunteer firefighters in the region who already give up so much of their time to help others, but still had time to lend me a hand.

Finally, thanks to my best friend Robbie Frame. Without his sleuth work, editing skills, and unfaltering love and support, this book would not have been published.

Chief Director's Message

In 2004, Halifax Regional Fire and Emergency (HRFE) celebrated the milestone of 250 years of organized firefighting in the Halifax Regional Municipality (HRM). This milestone holds great significance for our department, the oldest organized fire service in Canada.

Throughout World Wars One and Two, the Halifax Explosion, the Bedford Magazine fire, and many other major blazes and catastrophes, the fire service has always been there for the people of Halifax, Dartmouth, Bedford, and Halifax County.

Today, the fire service offers many diverse, high-tech rescue services. It has evolved from thirty-eight fire departments into a single, complex organization, which is responsible for meeting the needs of close to 360,000 citizens.

Halifax Regional Fire and Emergency is committed to giving our community exemplary service, while sharing the responsibility with our citizens to develop a safe environment.

Today, the service's activities extend far beyond emergency fire response. We are involved in community emergency preparedness; Urban Search and Rescue; developing relationships with other organizations such as the Safe Communities Foundation, Salvation Army, and Red Cross; as well as developing working partnerships with the other levels of government.

In the future, I see HRFE becoming more active in a civil response role. Over the last seven years, we have had front-line experience with natural and man-made disasters such as Hurricane Juan, White Juan, 9/11, and Swiss Air Flight 111. The continuing threat of terrorism that we face today may well demand the services of HRFE.

I would like to extend my heartfelt thanks to all members of the Halifax Regional Fire and Emergency who have played such a large part in protecting our communities. Their hard work and dedication ensures that we can continue to offer a world-class emergency service to the residents of the Halifax Regional Municipality.

Michael Eddy
Chief Director
Halifax Regional Fire and Emergency

Halifax Regional Fire & Emergency

Mayor's Message

ON BEHALF OF Halifax Regional Council, I am pleased to congratulate Halifax Regional Fire and Emergency on its 250th anniversary.

Fire suppression, prevention, and control have always been a necessary part of community life. All of us are reminded of the contribution that our fire departments have made in the protection of life and property throughout the years.

During the course of our city's history, many changes have taken place within the fire service; today we have a modern, well-trained and well-equipped department of which the citizens of Halifax Regional Municipality can be proud. This department has developed to its present state of efficiency through the efforts of many men and women in its 250-year history.

I am pleased to acknowledge that in its 250th year, Halifax Regional Fire and Emergency continues to play a vital role in protecting the citizens of Halifax Regional Municipality.

Peter J. Kelly
Mayor
Halifax Regional Municipality

A Firefighter's Prayer

When I am called to duty, God—wherever flames may rage

Give me strength to save a life, whatever be its age.

Help me embrace a little child—before it is too late

Or save an older person from the horror of that fate.

Enable me to be alert, and hear the weakest shout,

Quickly and efficiently to put the fire out.

I want to fill my calling, to give the best in me

To guard my friend and neighbor, and protect his property.

And if according to Your will—I must answer death's call

Bless with your protecting hand,

my family one and all.

Table of Contents

Map and Statistics of Halifax Regional Fire and Emergency......... viii
List of fire stations in the Halifax Regional Municipality ix

	Introduction 1	
Chapter 1	Firefighting in the Early Twentieth Century...............19	
Chapter 2	Fresh Starts: Development of Rural and Urban Fire Departments 52	
Chapter 3	Evolution of Firefighting in the HRM 87	
Chapter 4	Many Communities, One Department: The Impact of Amalgamation..................103	
Chapter 5	Noteworthy Fires and Disasters from the Last Thirty Years ..116	
Chapter 6	Then and Now 133	

Last Alarm:

Line of Duty Deaths......................146

Image Sources......................147

Bibliography......................148

Halifax Regional Fire and Emergency

Headquarters
● 62 Fire Stations

Area of Coverage: Approximately 5,577 sq. kilometres
Population: 360,000
1,259 Fire Fighters (career & volunteer)

Halifax Regional Fire and Emergency Statistics

Headquarters: Dartmouth
62 fire stations
Area of coverage: 5,577 square kms
Population: approximately 360,000
1,259 members (career and volunteer)

Fire Stations in the Halifax Regional Municipality

Station #1. Central Headquarters—Alderney Dr., Dartmouth
Station #2. University Ave., Halifax
Station #3. West St., Halifax
Station #4. Lady Hammond Rd., Halifax
Station #5. Bayers Rd., Halifax
Station #6. Herring Cove Rd., Halifax
Station #7. Knightsridge Dr., Halifax
Station #8. Convoy Rd., Bedford
Station #9. Metropolitan Blvd., Sackville
Station #10. Sackville Dr., Sackville
Station #11. Patton Rd., Sackville
Station #12. Windmill Rd., Dartmouth
Station #13. King St., Dartmouth
Station #14. Second St., Dartmouth
Station #15. Pleasant St., Dartmouth
Station #16. Caldwell Rd., Eastern Passage
Station #17. Cole Harbour Rd., Dartmouth
Station #18. Main St., Dartmouth
Station #19. Crowell Rd., Lawrencetown
Station #20. Lawrencetown Rd., Lawrencetown
Station #21. Lake Echo
Station #22. North Preston
Station #23. Chezzetcook.
Station #24. Musquodoboit Harbour
Station #25. Ostrea Lake
Station #26. Jeddore
Station #27. Owl's Head Harbour
Station #28. Sheet Harbour
Station #29. Moser River
Station #30. Highway #7, Tangier
Station #31. East Ship Harbour
Station #32. Mooseland
Station #33. West Quoddy
Station #34. Powers Rd., Tangier
Station #35. Cooks Brook
Station #36. Meaghers Grant
Station #37. Elderbank
Station #38. Middle Musquodoboit
Station #39. Upper Musquodoboit
Station #40. Lantz
Station #41. Waverley
Station #42. Wellington
Station #43. Grand Lake
Station #44. Fall River Rd., Fall River
Station #45. #2 Highway, Fall River
Station #47. Goffs
Station #48. Beaver Bank
Station #50. Hammonds Plains
Station #51. Upper Hammonds Plains
Station #52. Hatchet Lake
Station #53. Terence Bay
Station #54. Shad Bay
Station #55. Seabright
Station #56. Black Point
Station #57. Head of St. Margaret's Bay
Station #58. Timberlea
Station #59. Lewis Lake.
Station #60. Herring Cove
Station #61. Ketch Harbour
Station #62. Harrietsfield
Station #63. Sambro

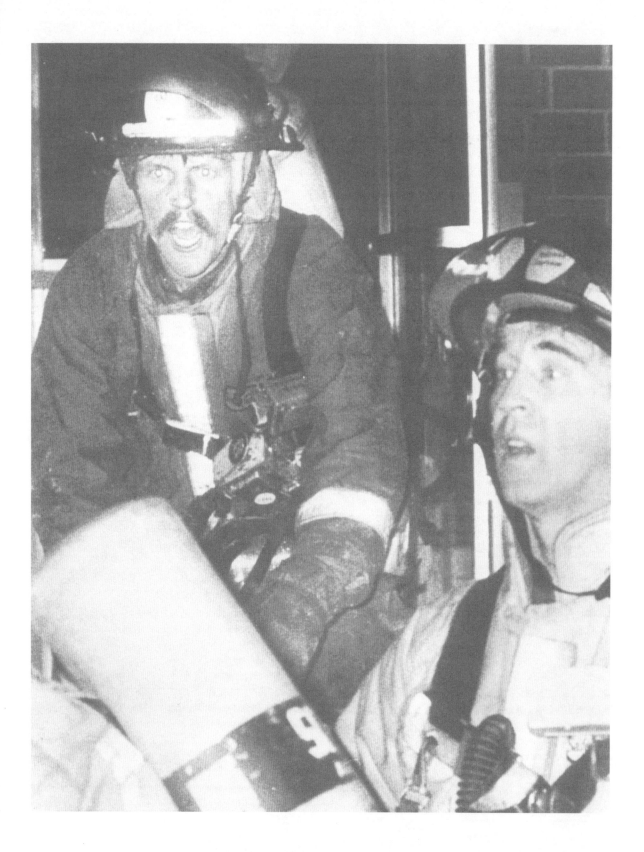

Introduction
1754–1900

A fire engine races by with its lights and sirens blazing. In moments of distress, this iconic image provides a sign of comfort and hope. Firefighters have long been admired by the communities they serve and it's not hard to see why. Whether paid or volunteer, urban or rural, when the alarm sounds, a firefighter's job is to rush to a scene that most of us would want to flee.

Firefighters respond to a call.

Long gone are the days of clanging fire bells, hot steam spewing from fire engines, and a firefighter taking great pride in the ability to "eat smoke" (it was once a source of pride to be able to breathe smoke without having to come out for air). Without today's modern breathing apparatus, legend has it that firemen used to grow long beards so they could soak their whiskers in water; when faced with a fire, they'd clench it between their teeth to act as a filter. Advances in technology have meant firefighters not only have better tools to work with, but also more protection and training in preparation for the hazards inherent to the job. Where a volunteer firefighter once had only muscle power to rely on when battling smoke and flames, today's firefighters are well-trained professionals who are fully prepared for the ring of the alarm.

Halifax's Firefighting Heritage

HALIFAX'S PROUD firefighting heritage spans 250 years; consequently, the city lays claim to many firsts in Canadian firefighting history, including the country's oldest fire department and fire insurance company.

Although firefighting can be traced to the founding of Halifax in 1749, the town's first organized firefighting group, the Union Fire Club, was officially established on January 14, 1754. Four years before the club was founded, the young town, which was little more than buildings under construction, was struck by a fire on July 11, 1750. It was believed to have been the first fire of major proportion in Canada and threatened to wipe out the entire community.

Firefighting was serious business for the club's thirty-four volunteer members. As the fire bells sounded in Halifax at any hour of the day, the firefighters rushed to the engine houses to run with the hose reels; or, if too far away, they ran to the scene of the fire, often in their best clothes. In the dead of night, the volunteer firefighters might be woken up by a military policeman, or night watchman rapping on their doors crying "FIRE... FIRE!"

Fourteen years after the founding of the Union Fire Club, the Union Engine Company, a larger and more sophisticated organization, was formed. This volunteer organization remained in service until 1894 when it was officially named the Halifax Fire Department.

Halifax is also where the first hand-propelled fire engine, the first steam engine, named *Victoria*, and the first motorized pumper, named the *Patricia*, were put into service.

In Upper Hammonds Plains the community laid claim to the only all-black fire department in the country. In 1964, the first meeting was held in the rural black area with the hope of establishing its own fire department. The volunteer department got off the ground two years later.

Early Equipment, Early Rules

IN 1749, firefighting equipment in Halifax included two fire engines; leather discharge fire hoses; three-gallon buckets for carrying water; axes and hooks for entering, and if necessary tearing down buildings; bed keys for disassembling wooden frame beds, which were in many cases a homeowner's most valuable possession, and bags and baskets for salvaging other household belongings. The rules for firefighters were clear and strict: "Every member shall be provided with two leather buckets marked with their names on one side and the Senior Fire Club on the other, and two bags each, made up of four yards of canvas. That said, buckets shall be hung up with the bags in some convenient place of each member's house to be ready for all emergencies."

Upon the cry of "Fire!", the members of the fire club were expected to immediately respond to the place of the blaze, with their buckets and bags. If a volunteer was not at a fire without a valid reason he could receive a fine.

The Union Engine Company

THE UNION ENGINE Company was formed on August 8, 1768. It may have been a re-restructuring of the Union Fire Club or something new. The early minutes of the company's meetings have not been handed down and may never be known for certain.

Members of the Union Engine Company in dress uniform stand beside No. 3 steam fire engine and the old hand fire-engine in front of Queen Street Engine House in Halifax, c. 1878

The Union Engine Company grew quickly as the community expanded. In 1790, fifteen firefighters were listed on the company's roll. The engine house was thought to be located on or near the Grand Parade, Halifax's historic centre. By 1812, the company had twenty-eight volunteers in three divisions. In 1826, membership was sixty and each new fireman had to pay an eight-dollar initiation fee.

While there are tales of rival fire companies ending up in brawls and gun fights in other parts of Canada and the United States, there is no record of this happening in Halifax. The rough tactics employed by competing firefighters as depicted in Martin Scorsese's 2002 film, *Gangs of New York*, don't seem to have taken place in Halifax.

Fire Wards

Leather bucket used to carry water to a fire, 1750

IN 1762, a local statute provided for the appointment of fire wardens (also called fire wards), and by 1812, Halifax was divided into twenty small fire districts, each under the jurisdiction of a court-appointed fire ward. A fire ward's duties ranged from inspecting chimneys to supervising the formation of bucket brigades to co-ordinating the removal of personal belongings from burning buildings. Fire wards were easily identified, as they wore a badge and carried a six-foot-long red staff topped with a six-inch brass spear tip.

In the 1800s, the responsibility for fire protection services was of tremendous importance. Among the more prominent individuals who served as fire wards were Samuel and Edward Cunard, Alexander Keith, William Stairs Sr., William Stairs Jr., John Leander Starr, Andrew Mitchell Uniacke, Benjamin Weir, and William Young.

Fire wards were responsible for administering the laws against the firing of rockets and fireworks and for licensing chimney sweeps. Homeowners who didn't have their chimneys swept regularly had to pay a fine. Fire wards also had the power to decide if adjacent houses should be torn down in an effort to prevent the spread of a fire.

Halifax had an especially large number of protection or salvage companies, according to fire historian Donal Baird. The Hand in Hand was established in 1789, and other similar companies, the Phoenix and the Heart in Hand, came into existence soon afterwards. While the city supplied buckets, ladders, and most likely fire engines, protection companies supplied their own salvage bags, baskets, and tools for the recovery of goods.

Probably the first ladder company in Canada, Halifax's Axe Company was founded on July 12, 1813. Arriving at the scene of a fire, the company would cut away burning parts or pull down a neighbouring structure to prevent the spread of flames.

Besides more formal organizations, many local residents kept two or three buckets made of leather or wood at the ready in case of fire. When an alarm sounded, volunteers would form two lines—one carrying water to the fire and the other returning empty buckets to be refilled.

A news article describes a typical Halifax fire scene from the early 1800s: "The sloshing buckets passing hastily from hand to hand, the perspiring leather caped gentlemen at the engine, bowing to each other alternately like clockwork toys as the hand bars went up and down, the paltry stream of water that gushed and sank with the movements of the pump, the gentlemen of rival fire companies watching in case 'one of ours' should catch fire from 'one of theirs.'"

The most important piece of firefighting equipment at the time was the hand-pumped fire engine, and in 1826 Halifax had four in operation. Bucket brigades fed these early fire engines with water either from strategically located tanks or cisterns, or the harbour.

As they do today, early fires drew crowds of curious bystanders. But unlike today, spectators were sometimes forced to relieve exhausted firefighters by working the hand engines. In 1835, fire wardens complained about these often unco-operative "volunteers." To appease the wardens, the House of Assembly created a special group of three hundred so-called "engine workers", whose sole duty was to man the pumps, which provided a stream of water sprayed from a goose-neck nozzle on the end of a short length of hose. Firefighters were constantly in danger as the short hoses meant they had to fight fires at close proximity. The hoses were so short that several times fire engines went up in flames.

In the mid-1800s, hand-pumped fire engines were being replaced by steam engines. According to Mike Parker's *The Smoke-Eaters,* bucket brigades remained in existence right up until the early twentieth century.

Military Firefighters

HALIFAX was lucky to have both a naval dockyard and an army garrison, each of which had its own fire engines. In 1800, there were eight thousand civilians in Halifax and four thousand military personnel. But the goodwill of the military was sometimes tried when they had to provide manpower because turnout of civilian volunteers was poor. On one occasion, reported on October 26, 1837, soldiers working under a civilian fire warden were reported to have yelled, "Why don't you get your own men!" Twenty years later, while many of the town's firefighters were away celebrating New Year's Day, the Halifax garrison again "saved the day" by fighting a twenty-two-hour fire at St. Matthew's Church on Prince Street.

Fire Insurers

Halifax Fire Insurance Association, 1909

FIRE INSURANCE in Nova Scotia can trace its roots back to 1784 when the Phoenix Fire Insurance Company of London "accepted a risk" in Halifax. It was in their interest to promote fire protection in the city, and donations of money or equipment were often made in the hopes of strengthening fire brigades. For example, the Halifax Fire Insurance Association provided a fire engine to Halifax in 1827, as well as two hundred pounds toward building a water reservoir. In the early part of the 1800s, the Halifax Insurance Company was formed, becoming the first Canadian fire insurance company.

An article which appeared on July 26, 1828, in the *Acadian Recorder*, explains another way insurance companies supported firefighting: "It has been the practice of the Halifax Insurance Company of this place to give the first engine arriving at a fire a fee of $7… Such rewards are highly valuable and praiseworthy." While rewards probably improved response times, they also generated conflict between civilian and military fire departments. Civilian companies were especially bitter because the military had an unfair advantage, as for years military buglers sounded the alarm. The Halifax Insurance Company would eliminate engine bounties by 1849.

Fire Alarms

IN 1822, a fire bell was imported from London and installed at the market opposite the town guardhouse. A larger, 417-pound bell was presented by the Sun Fire Company to Mayor Alexander Keith in 1844 and installed at the central engine house at the Grand Parade. It is believed that this bell replaced army buglers as the town's primary fire alarm.

Dartmouth businessman John P. Mott bought this Perry fore-and-aft hand fire engine in 1844 to protect his chocolate factory and spice mill from fire.

The City's Firefighting Record

WHILE FIRES were destroying large portions of major cities throughout North America in the mid-1800s, Halifax experienced nothing of the kind, according to B.E.S. Rudachyk's paper, "At the Mercy of the Devouring Element":"Halifax, in comparison with her North American counterparts, was not ravaged by the tyrant flame. From 1830 to 1850, although Halifax firefighters contended with 281 fires, only eleven of these destroyed five or more houses per incident— and only one more than thirty...This represents one of the finest firefighting records, for contemporary cities of comparable size, anywhere in North America. It is a remarkable record considering that during the 19th Century any uncontrolled fire— however trifling its initial appearance—could, and regularly did, destroy from five to twenty buildings."

Halifax would have several major fires in the second half of the nineteenth century but even then, it would not experience fires on the scale of other cities like Saint John and Quebec. Though Halifax had some stone buildings, wooden buildings with shingled roofs were predominant. The town had one advantage in that its streets had been regularly laid out to be fifty- to sixty-feet wide. But Rudachyk credits Halifax's success with what he called "the effective organization and zealous co-operation of the fire establishment's personnel." The efficiency of the force was evident during a mid-1840s emergency— fourteen fire engines were on the scene within ten minutes of an alarm for a fierce waterfront blaze, and axemen rapidly cut down a nearby wooden building that might have spread the fire.

However, firefighters didn't always receive rave reviews for their work. After a New Year's Day fire in 1857, the Halifax Water Company, formed in 1844 as a private company, was criticized for not providing city firefighters with an adequate supply of piped water. During a Hollis Street fire in 1858, the Ladder Company was criticized for its tardiness and the Union Engine Company for a perceived "lack of direction."

Halifax didn't escape all major fires. A barracks fire on December 24, 1850, destroyed about one hundred houses. The Granville Street Fire of September 9, 1859, proved to be an even worse disaster—four acres of the city's centre

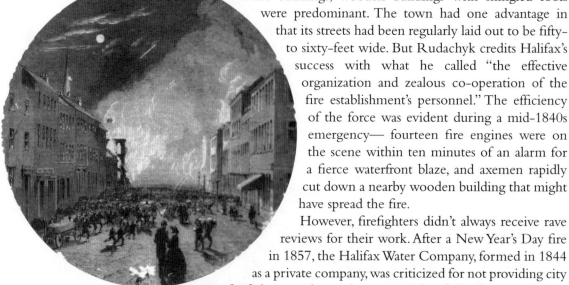

Granville Street Fire,
Sept. 9, 1859

were destroyed. On October 15, 1859, *The Illustrated London News* reported that:"Houses and stores, wooden, brick and stone, all alike fed the flames, until, of the two extensive blocks touching on Hollis and Barrington streets, with Granville-street (the Haligonian's paradise), running betwixt them, nothing escaped except one store." Several lives were lost, serious injuries were sustained, and damages topped three-quarters-of-a-million dollars. Sixty of the city's finest buildings were in ruins. Over the next two years, brick and stone structures replaced what had been wooden buildings.

Two years later, on January 12, 1861, a fire struck in freezing weather, destroying an area that included most of George and Prince Streets. In total, twenty-two offices, four dwellings, twenty-two stores, and the Halifax Library were reported damaged.

The First Steam Engine—Dissension in the Ranks

SINCE MAJOR FIRES were no longer strangers to Halifax, something had to be done—council decided to take action. After fire insurers threatened to withdraw fire coverage, council passed regulations to create a brick and stone district and expropriated water service for municipal control. They also set aside money for a steam-powered fire engine; Industrial Age firefighting equipment was on its way. The mayor and an alderman went to Boston to make the purchase.

An Amoskeag engine was purchased from Manchester Locomotive Works in Manchester, New Hampshire, in March 1861, and was named *Victoria* in honour of England's long-serving Queen. A second steam engine was added

Victoria, Canada's first steam engine, c. 1861

The steam engine *Albert*

soon after and dubbed *Albert* after the Queen's husband. In 1862, Halifax boasted that it had "two first-class steam engines," five hand engines, seven hose reels, eight thousand feet of hose and an assortment of equipment that was "all in prime order and of the best material and style."

The steam engines that arrived in Halifax in the early 1860s profoundly changed firefighting, and the city was more prepared for fire than ever before. However, the rapid modernizations prompted the resignation of many Halifax firefighters. Dissention rose in the ranks, and members of the Union Engine and Axe fire companies left their jobs. According to author Mike Parker, many firefighters saw the steam engine as a threat to their existence. A machine could produce more powerful streams of water, didn't need to stop to rest, and could perform the work of many firefighters. Council didn't have any difficulty finding new volunteers to replace the firefighters who had walked off the job, and a new volunteer engine company of about 150 men was formed. Ex-members formed the Union Life and Fire Protection Company (later the Union Protection Company), and acted as a salvage and auxiliary force. For several years there was a bitter rivalry between the new volunteers and the Union Protection Company, but eventually wounds healed, rivalries faded, and a friendlier relationship emerged.

Sutherland & Craig hose reel, 1887

The Truro company Sutherland & Craig built hose reels in the 1880s. J.L. Sutherland (at left) and S. Craig pose beside one of their products decorated for a Natal Day celebration.

Hose Reels

FIRE HOSES in the latter 1800s were heavy. A one-and-a-half or two-inch double-riveted leather hose, with brass couplings and iron hand rings for pulling, could weigh eighty pounds. Because of their weight, hose reels were needed to carry the hoses. A typical nineteenth-century two-wheeled, hand-drawn hose reel carried five hundred feet of hose or more and could weigh six hundred pounds when fully loaded.

Even though they were manufactured, some firefighters continued to design and build their own hose reels and carts right into the twentieth century. The Lawrencetown Fire Brigade built two-wheeled hose carts in 1917 with boxes added to hold axes and other firefighting equipment. Initially hand-drawn, the hose reels were later drawn by horses in four-wheeled carts and wagons. Eventually they were towed behind automobiles.

By the late 1880s, lower-maintenance rubber hose was replacing the copper-riveted leather kind, which had a tendency to dry out and crack if not regularly oiled. Hose got even better with the development of unlined and rubber-lined cotton hose, which tended to be lighter, stronger, and easier to store.

South Street Poor House Fire

THE CONCEPT of the Victorian workhouse, or Poor House, as it was called in Halifax, has passed from our collective memory and is now conjured up chiefly by Charles Dickens' vivid recreations of life in a workhouse in his classic novel *Oliver Twist*. In Halifax, thirty-one people died on the night of November 6, 1882, when the South Street Poor House burnt, leaving just a shell behind; more lives were lost in that blaze than in any other fire in the city's history. The fire, which started accidentally in the kitchen area of the building, was discovered about 11:30 P.M. It spread quickly through the ceiling to the walls and elevator, and from the elevator shaft to the rest of the massive, fortress-like structure.

Remains of the Poor House Fire, Halifax, Nov. 7, 1882

Many of the 343 residents were hysterical when the city firefighters arrived. When the main door was knocked down with a fireman's axe, there was pandemonium. A crowd of people streamed outdoors—mothers nursing infants, some fifty to sixty children, elderly women and feeble old men. Some were partially clothed, others wrapped in blankets, others naked. Police, firefighters, clergy, reporters and spectators rushed into the building to assist or to carry the blind and handicapped to safety. Suddenly, the building erupted into a raging inferno, with forty or fifty patients still trapped in the infirmary on the fifth floor.

The November 7 issue of the *Citizen and Evening Chronicle* described the gruesome scene: "The fire was fiercely burning bright in the hospital, and above it heat cracked the roof until molten lead poured down in streams of brilliant fire and slates flew on every side in a deadly shower, rendering any near approach to the building almost certain death....Far above the roar of flames and crack of bursting slates were heard the cries of the wretched patients in the hospital, who were roasting to death. Most of them...were helpless, [and] could not leave their beds...."

An attempt was made to raise ladders to reach patients but they proved to be too short. After a fireman was knocked down by falling bricks, firefighters had to stop their rescue attempt and resign themselves to witnessing the tragedy. An inquest into the fire would later reveal the incompetence of poor house managers during the fire and expose the fact that the building had not been properly fireproofed.

The Formation of the Halifax Fire Department

IN THE MID-1800s volunteerism still played a key role in providing city services in Halifax, due to council's limited powers of taxation and borrowing. One of many voluntary civic services was the fire brigade. Volunteers were attracted to firefighting for excitement, a sense of responsibility, desire for social interaction, or as a way to get an exemption from serving in the militia or on a jury.

In 1884 firefighting still operated under a volunteer system. The city's annual report for that year states: "Halifax is still old-fashioned enough to prefer her extremely effective fire department composed of volunteers, to a paid department as now exists in most other cities. So long as such splendid bodies of men as those forming the Union Engine Company, the Union Protection Company, the Union Axe and Ladder Company are ready and willing at a moment's notice to combat the devouring elements, the citizens of Halifax should feel profoundly grateful to their faithful and brave men for their self-denying and arduous services for the common good."

The Union Engine Company continued as a volunteer organization until 1894, when it was reorganized as the Halifax Fire Department. It then started to operate on a part-paid, part-call basis. John Connolly became the new department's fire chief. Like most changes, it wasn't a smooth transition, at least according to predictions made by the *Daily Echo*. To quote a November 2, 1894 article, "There is likely to be some little trouble between some members of the Union Engine Company and the new department."

Nine volunteers formed the Dartmouth Fire Engine Company in September 1822, under Captain William Allen. Before 1822, members of the Halifax Union Engine Company fought Dartmouth fires, but a lengthy travel time did not always result in very effective firefighting. After the Dartmouth Fire Engine Company was established, mutual aid between the two departments became possible. For example, Halifax firefighters supported their Dartmouth colleagues during the Stanford Tannery fire of July 1844 by ferrying a fire engine across the harbour.

The Axe and Ladder Company

In 1865, the second branch of the Dartmouth department, the Axe and Ladder Company, was formed. As the town expanded it became apparent that more firefighters, and more modern equipment, would be needed. In 1876, the Union Protection Company was established as the third branch of the department, with a primary purpose of saving citizens and moveable property in a fire. Initially, the new company had, at most, fifteen members, each of whom provided their own personal equipment. The company also had a donated horse and wagon which was used to haul equipment to fire scenes.

The engine house was on Queen Street, a two-and-a-half storey wood frame building with a pediment gable roof. A more spacious fire station on King Street later replaced the Queen Street location. In 1878, a new steel fire bell, weighing 870 pounds was placed in the station's tower.

The Union Protection Company's first major fire was at the Hartshorne and Tremaine gristmill in Dartmouth Cove in the spring of 1878. The large structure burst into flames, lighting up the whole downtown area. It was reported that the windowpanes of houses on nearby Portland Street became too hot to touch. Firefighters worked through the night evacuating residents, removing furniture and personal belongings from their homes, and saving merchandise from nearby shops.

Later that year, the company would assist at a fire that destroyed Oland's Brewery. While there aren't accounts of firefighters receiving beer to keep them going during the long, hot fire, the press did report at a different brewery fire that all furniture and stock were lost, with the exception of "ten casks of spirits which were got out."

In 1878, the Dartmouth Engine Company, whose firefighting record was now among the best in the country, bought a brand new horse-drawn steam fire engine from the Sibley Co. of New York. It replaced a hand-fire engine that had been acquired in 1872. Called the *Lady Dufferin*, after the wife of the Marquis of Dufferin and Ava, it was considered "one of

the most efficient machines in Eastern Canada." The engine remained in service fighting fires until 1919. Its days ended only after combustion engines made horse-drawn equipment undesirable.

Dartmouth Axe and Ladder Company, c.1860

Colin McNab, the owner of McNab's Store in Dartmouth, was associated with the Dartmouth Union Protection Company from 1884 until his death in 1948. His store was a popular place for storytelling, and on a winter evening he listened to one story about the situations firefighters faced in the early days. The story was retold by Syd Gosley, a former director of the Dartmouth Heritage Museum:

"A certain man in Dartmouth, no name mentioned, suffered the annoyance of a smokey chimney and made inquiries as to what he could do to cure the problem. He was told that gunpowder in sufficient quantities would remove all obstructions. He went home with half a pound of powder quite pleased with himself. He never told his wife or any others of his large family about his intentions, all he said was he had a surprise in store for her. 'Poor woman,' she'll never forget it. He got his supper and then while everyone's back was turned he dropped the can of powder into the kitchen stove. In about 4/5ths of a second those people in the house were so mixed up that they didn't know one from the other and all were shouting 'dynamite!' and if they caught the so and so…. Then they commenced to look for

The fire engine *Lady Dufferin*, Dartmouth, c. 1900

Firemen with the Dartmouth Axe and Ladder Company, 1911

each other, some were in the cellar, others out in the yard next door, and a couple of kids had taken to the trees opposite. The house could not be repaired for less than it had originally cost to build, and he who tried the drastic remedy went about with his arm in a sling."

In 1932, the Dartmouth Engine Company, Axe and Ladder Company and Union Protection Company united to form the Dartmouth Fire Department. In 1961, with the amalgamation of the outlying areas around Dartmouth, the Woodside Fire Department would become part of the Dartmouth's department.

A Proud History

THROUGHOUT the many events and disasters of the last 250 years, such as World Wars One and Two, the Halifax Explosion, the Bedford Magazine fire, and countless other major blazes and catastrophes, firefighters have always worked faithfully for the residents of Halifax, Dartmouth, and surrounding communities. When municipal amalgamation took place in 1996 and the former cities of Halifax and Dartmouth, and several surrounding towns and communities, merged to become the Halifax Regional Municipality, the Halifax Regional Fire and Emergency Service was officially born.

Dartmouth Fire Station on King Street, c. 1920

Halifax
Fire Department
Aerial Truck
c. 1900

Chapter 1

Firefighting in the Early Twentieth Century

The fifty-year period from the turn of the century through to the post-war years was a time of tremendous change. The two World Wars had a huge influence on the port city of Halifax, which was Canada's major military centre during World War One and the country's main naval port during World War Two. At the outbreak of war in 1914 there was an overwhelming patriotic response in Halifax, which saw the departure of a reported 284,455 personnel. Canadian, Commonwealth, and American soldiers all departed from the hub of Halifax. During World War Two, the city saw more than one hundred and fifty troop convoys embark.

Halifax Fire Chief John Connolly, appointed in 1894

Grim history was also made in Halifax in 1917 when the Halifax Explosion, the most devastating man-made explosion before the dropping of the atomic bomb in Japan, rocked the city. A second but not as serious explosion in 1945, the Bedford Magazine Fire, was also a highly damaging event.

The period also saw new volunteer fire departments spouting up outside Halifax and Dartmouth. Among the departments formed during this half-century were in Bedford and Black Point.

Soon after the new century rolled in, Halifax's fire chief, John Connolly, who had been appointed in 1894, was replaced in 1903 by Patrick Broderick. Chief Broderick remained in office until his death in December 1916. As a job perk, the fire chief was provided with a horse-drawn buggy to help him travel between the eight engine houses and respond more quickly to emergency calls. At the front of the chief's buggy was a gong, a standard feature on all early fire apparatus, which served to warn pedestrians and traffic to clear the street for oncoming firefighters. The only piece of equipment carried on the chief's wagon was a first-aid kit for both men and horses. During Broderick's first year as fire chief, property losses from fires in Halifax were reported to be $135,224. By 1914, they were up to $315,710 and by 1920 they were reported to be $648,074. This dramatic increase reflects the growth of the city and the fact that the fire department was becoming more organized and therefore keeping better track of statistics.

Halifax Fire Chief Patrick Broderick and a driver, c. 1910

An invaluable piece of equipment in the early 1900s for Halifax firefighters was the ladder. Wooden ladders were the norm until the mid-1900s when metal started to be more widely used. Firefighters relied on ladders to get to the upper stories of buildings, rescue people, and save property. By 1849, Halifax owned two hook-and-ladder carts; each carried hook ladders and wall ladders twelve metres long, support poles, and hooks and chains for pulling down walls and roofs off burning buildings. By the early 1990s, Halifax owned 230 metres of ladder with forty-five metres in reserve.

One of the perennial problems for the fire department at this time was Halifax's water service, which was taxed by excessive waste. The estimated daily water usage in 1908 was an unbelievable 1,323 litres for every citizen, according to *Halifax: The First 250 Years*. Today, the average person uses about 454 litres of water at home every day. In 1913, the construction of a water tower on Shaffroth's Hill at the north end of Robie Street equalized the pressure for consumers in that part of town, making the system more efficient and reliable, and consequently reducing water waste.

The Acadia Sugar Refinery

In Dartmouth, on February 1, 1911, at about 5:20 P.M. fire broke out in the Acadia Sugar Refinery warehouse, located on the dock in Woodside. Flames quickly engulfed the building. High winds fanned the flames which spread from the dockside warehouse to the main building where the sugar was stored. Sugar, which is highly combustible, added to the fire's intensity. At the time of the fire, there were fifteen thousand barrels, forty thousand bags of raw sugar, and ten thousand dollars' worth of refined sugar in bags.

The refinery fire brigade was the first on the scene, quickly followed by brigades from Woodside, Dartmouth and later Halifax. Unable to save the building, the teams did save a nearby powerhouse. Tragedy could not be averted, however, as the fire killed a fifty-eight year-old man named Thomas Henneberry who became trapped in the warehouse. The fire loss was estimated at $1.5 million and left 150 people unemployed. The refinery was rebuilt in 1914.

Halifax's Gamewell fire alarm system

THE EARLY 1900s not only brought advances in firefighting equipment but also more sophisticated means of alerting firefighters when fire broke out. In the early 1860s, Halifax installed a telegraph fire alarm system for a cost of ten thousand dollars which used a series of signal boxes wired to an engine house. In 1902, that system was replaced by a Gamewell telegraph alarm, which was operated through one hundred and four kilometres of live wire and one hundred and thirteen boxes. Three linemen were employed to take care of the system. Halifax firemen were generally called to a fire by a policeman on duty who was also often the person to have pulled the alarm. During electrical storms, however, the bells were often set off and firefighters would run in all directions in search of a non-existent fire. According to historian and novelist Thomas Raddall, the system worked in the following way: "The alarm was sounded over the city…ringing the location by a code which every housewife kept posted on her kitchen door."

In 1945, the city got a new Gamewell system. It was located in a brick building on Summer Street and served the city until the early 1990s. Forty new alarm boxes (in addition to the 150 that already existed) were installed throughout the city. A coded series of printed-out taps gave the signal box's location to the alarm headquarters and engine houses simultaneously. Alarms that were phoned in first went to headquarters and then were relayed using a transmitter wheel (as shown on the table in the photo).

Halifax's fire alarm system did have shortcomings which sometimes led to disastrous consequences. A box alarm failure on February 14, 1926 prevented adequate response to a $94,000 fire at Ben Moir's Limited on Pepperell Street. In a report to the fire wards committee less than a month later, it was revealed that the first information the fire department had received about the blaze was from a person who notified the Quinpool Road station that Ben Moir's was "all on fire."

After receiving the notification, several attempts were then made by the station to transmit an alarm, to no avail. Instead, other nearby stations had to be notified by phone. During the time it took for the other stations to arrive, Captain Harber and Hoseman Gorman arrived at the scene and attempted to haul a line of hose into the building, getting overcome by smoke in the process. Harber was able to get out, despite his unsuccessful attempt to rescue Gorman.

When firefighters from West Street station arrived, Harber told them that Gorman was in the building. Hoseman J. C. Doherty immediately broke a second-floor window and entered the burning building. Eventually Gorman was found and carried out but he was unconscious and bleeding from his nose and mouth. Attempts were made to resuscitate him before he was taken to the Victoria General Hospital. He died shortly after. Gorman, who had been with the department since 1919, was called a trustworthy and faithful fireman. Harber was also taken to hospital, but was later released.

Canada's first motorized pumper, *Patricia*

L–R: Chief Edward Condon, Comptroller Hines, Deputy Chief William P. Brunt, Capt. John Brommit, unidentified, unidentified, Hosemen William Connors and Ned Strachan, driver Billy Wells, Claude Wells (Chief Condon's driver), Hosemen Frank Killeen, Art Sheehan, Joseph Ryan, and Walter Hennessy

DURING CHIEF BRODERICK'S term in office he brought in the department's first motorized pumper. Not only a milestone for the department, it was the first of its kind in Canada. The *Patricia*, a 1912 American LaFrance type twelve triple-combination pumper, chemical engine, and hose wagon, arrived in Halifax on March 13, 1913. Named in honour of the governor general's daughter, the pumper cost $10,800 and came with all the latest equipment. This included a six-cylinder, seventy-three horsepower motor, a five hundred gallon-per-minute rotary water pump, a thirty-five gallon chemical tank, and twenty-two feet of four-and-a-half-inch and two-and-a-half-inch stiff suction hose. The hose had suction Siamese and reducers. The pumper was also equipped with divided hose body for twelve hundred feet of two-an-a-half-inch fire hose and two hundred feet of chemical hose.

Halifax didn't have the *Patricia* long before it suffered $7,500 in damages during the Halifax Explosion on December 6, 1917. Four of the men featured in the photo—Condon (who replaced Chief Broderick as chief of the Halifax Fire Department in January 1917), Brunt, Killeen and Hennessy—were among nine firemen killed in the explosion. The pumper was rebuilt by LaFrance in Elmira, New York, and returned to duty within a few months after the explosion. It remained in service until 1942; three years later it was put on the auction block. "The 'Daddy' of Halifax fire pumpers, which led in the fight against most of the city's major fires for more than three decades, has pulled down its battle flag at the end of its days of action—and now nobody wants it….The scrap pile will likely be the answer," the *Halifax Herald* reported on February 26, 1945.

On the clear, cold morning of December 6, 1917, the munitions ship *Mont Blanc* collided in Halifax Harbour with the Belgian ship *Imo*. The *Mont Blanc*, a French freighter, was carrying an explosive cargo of thirty-five tonnes of flammable benzine in barrels lashed to its deck, as well as 2300 tonnes of picric acid and two hundred tonnes of TNT in its holds.

The *Mont Blanc* had reached the outskirts of Halifax Harbour on December 5, having travelled from New York. It lay anchor and was awaiting permission to enter the Bedford Basin the next morning. Meanwhile, the *Imo* was anxious to get underway for New York to pick up relief supplies for Belgium.

On the following morning of December 6, the *Mont Blanc* made its way toward the Narrows, a constriction in the water at the mouth of the Bedford Basin. Following the necessary procedures, the *Mont Blanc* kept to its starboard, or Dartmouth side of the harbour, so that outgoing ships could pass on its port, or Halifax side. Unfortunately, at that same time the outgoing *Imo* was also on the Dartmouth side and on a collision course with the *Mont Blanc*. The ships exchanged warning whistles before the captain of the *Mont Blanc* ordered his ship to veer toward port. About the same time, the captain of the *Imo* ordered his ship to veer toward starboard. The *Imo* then reversed its engines, throwing its bow into the bow of the

This photograph, taken after the explosion, appeared on the front page of the Dec. 22, 1917 edition of the *Montreal Standard*. The *Imo* is shown at left.

Mont Blanc. Sparks flew, igniting the benzine spilling from ruptured barrels into the holds of the *Mont Blanc.* By about 8:45 A.M. the *Mont Blanc* was engulfed in flames that sent plumes of dark smoke into the sky. The ship then drifted toward Pier 6 in the city's North End, threatening to set it on fire. Realizing he could not control the fire, the ship's captain ordered his crew to abandon the vessel.

Unaware of the deadly cargo onboard the ships, hundreds of people gathered at their windows, rooftops, nearby streets, and at the top of Fort Needham to see the blaze. At the West Street fire station, an alarm sounded shortly before 9 A.M. The firefighters immediately knew that it was from dockyard Box 83 at Pier 6. It had become almost a daily routine to get such a call as the dock was continually catching fire as coal embers were dumped from ships' boilers.

The Box 83 fire call would have been routine had it not been for Constant Upham. Upham owned a North End general store and was among the few residents in the area with a home telephone. He could see that the fire aboard the *Mont Blanc* was more serious than burning embers, and phoned all the nearby fire halls. Firefighters from West Street, Brunswick Street, Gottingen Street, and Quinpool Road responded to his call. Despite the large response, the firefighters were limited in what they could do. The explosion was bigger than anything they had faced before.

Homes shattered in the deadly explosion

The *Patricia* suffers extensive damage

ON DECEMBER 6, 1917, when driver Billy Wells pulled the *Patricia* out of the station, Captain William Broderick, Captain Michael Maltus, Walter Hennessy, Frank Killeen and Frank Leahy, were onboard. They had to leave one on-duty firefighter behind at the station; when the alarm came in, he was sick with the flu in the bathroom and despite the chief's disciplinary threats, wouldn't come out. Captain Maltus then took his place. West Street firefighters were the first to arrive at the Pier 6 fire. For all but one of them, it would be their last alarm.

As the *Patricia* raced along Gottingen Street, Albert Brunt, a call fireman with the Halifax Fire Department and a house painter by trade, tried to jump on. But the part-time firefighter, who was pushing his paint cart along the street when he heard the alarm sound, couldn't get a secure grip on *Patricia*'s rails and slipped off, scraping his knees and hands. The firemen on the truck hollered after him as they headed towards the Pier 6 fire, but Brunt was a lucky man. He survived the explosion.

When the firefighters arrived at the pier, the heat was so intense they had to turn away. Chief Condon pulled Box 83 a second time to get additional help. Hose was taken from the *Patricia*, and a line was run to the pier. Wells moved the *Patricia* into position at the nearest hydrant. Then, just before 9: 05 A.M., only minutes after the sounding of the first alarm, the *Mont Blanc* exploded with a force larger than anything humankind had created before the United States dropped the atomic bomb in 1945 on Hiroshima, Japan.

Roof collapses on the North Street Station, 1917

It was five minutes after nine,
As those alive can tell,
That the beautiful city of Halifax,
Was given a taste of Hell.
—Unknown, cited in Paul Erickson, Historic North End Halifax

THE CREW OF the *Patricia*, except for Wells and Leahy, were instantly killed. Wells was thrown from the driver's seat of the vehicle, still clutching half of the steering wheel. His right arm and eye were badly injured. Moments later a tidal wave carried him up and back down Richmond Hill. He got tangled up in telephone wires and almost drowned. Miraculously, he survived.

Both Chief Condon and Deputy Chief Brunt were killed. John Spruin, a retired and well-respected fireman, had heard the second alarm and pulled on his fire suit. While driving a horse-drawn pumper along Brunswick Street, he was killed by flying shrapnel from the *Mont Blanc*.

Almost all of Richmond was destroyed, and much of the surrounding North End was devastated. The force of the explosion levelled buildings, crushed and trapped people, and overturned stoves, igniting fires. Shattered glass from blown-out windows became a deadly weapon. The explosion caused a tidal wave to sweep up Needham Hill as high as 18.3 metres. On its descent, it carried people and wreckage back into the harbour.

Halifax Fire Chief
Edward Condon

THE MORE THAN THIRTY permanent city firefighters and 120 volunteers who survived the explosion worked valiantly to douse the wooden houses on fire. Comptroller Hines commanded forces in the south section of Richmond, believing that Chief Condon was doing the same in the north. He didn't learn of Chief Condon's fate until later that night. Firefighters worked their way from fire to fire without the help of the *Patricia*, the department's only piece of motorized equipment. Accordingly, they turned to their dependable horses to haul their equipment. They were successful in fighting a fire at the west gate of the Wellington barracks (now Stadacona), and were able to stop the flames from spreading to southern sections of the city.

News of the explosion spread quickly. Within hours, trains carrying firefighters, apparatus, and hose arrived from New Glasgow and other parts of the province. However, the hose connections were not standardized and many could not be connected into Halifax's system. To add to the city's troubles, a fierce winter storm was blowing, dumping just over forty centimetres of snow on Halifax, and severely hampering rescue efforts. Novelist Hugh MacLennan described the scene the morning after the explosion in his book *Barometer Rising*:

"When dawn broke, Fort Needham looked like a long whale-back, pocked with hundreds of hummocks gray with sooty snow. But a fume of steam rose from the whole of it and blew southwest in the gale, and the thin line of men working methodically across it appeared like the vanguard of an attacking army stopped in its tracks digging in under fire."

Dominion Textiles Factory crumbles, c. 1917

STATISTICS VARY on the destruction and the number of people killed in the explosion. One count put the number of deaths at two thousand, with sixteen hundred destroyed buildings, six thousand homeless people, and twenty-five thousand people with damaged homes. There was an estimated $35 million worth of property damage.

On December 6, 1992, the seventy-fifth anniversary of the explosion, a black polished granite monument was unveiled at a ceremony at the fire station on Lady Hammond Road in memory of the nine firemen killed on that disastrous day. Every year since then, the fire department's fire honour guard has held a ceremony at the memorial in honour of the men and all firefighters around the world who have died in the line of duty.

Less than two weeks after the explosion, on December 17, 1917, John Churchill was appointed head of the Halifax Fire Department, replacing Chief Condon. He took over a department that was part paid and part call, with thirty-five permanent men and eighty-eight call men totalling 123 members. Churchill changed the department to a full-paid system, leaving behind the old volunteer system that had been in place since 1894. In 1919, there were eighty-six permanent men, paid a total of $129,549 in annual salaries. Each station had fourteen men on permanently with a hose, pumper and ladder. Problems arose during meal times, when half of the men would be off-duty and there weren't enough bodies to man the equipment.

Fireman Billy Wells

IN 1920, Halifax firefighters were paid twenty dollars a week and had only one day off in eight. Six years later, on May 26, 1926, Halifax firefighters became unionized, forming Local 268 of the International Association of Firefighters which was based in Washington, DC. As secretary of Local 268, Captain James Cody of the Halifax Fire Department fought with city council for more pay and better working hours. In 1927, firefighters won the right to have one day off in three during summer months. The following year, the department went on a two-platoon system and an eighty-four-hour work week. This was later followed by a three-platoon, seventy-two-hour week, and in 1955, this changed to twenty-four hours on duty, twenty-four hours off duty, with one free shift every eight working days. Today, firefighters in the core of the region work a twenty-four hour shift followed by seventy-two hours off. In the rural areas, career firefighters generally work from 8:00 A.M. until 5:00 P.M. on weekdays.

At a committee of the fire wards meeting held on June 13, 1919, less than two years after the Halifax Explosion, the fire chief read a report about the well-being of Wells, the former driver of the *Patricia,* who had been seriously injured in the explosion. The chief reported that Wells had received full pay from the fire department and compensation from the Relief Commission while he recovered from his injuries. Although one of his arms lacked muscle tone and remained somewhat handicapped, he was eventually able to return to work with the department as a special constable. He retired from the department in 1926.

"The first thing I remember after the explosion was standing quite a distance from the fire engine," Wells, then eighty-seven years old, told the *Chronicle-Herald* in an interview in December, 1967. "The force of the explosion had blown off all my clothes as well as the muscles from my right arm."

He described the gruesome sight of seeing dead bodies hanging out of windows, some without their heads. He also recalled crossing paths with two small children who were wandering alone in the street, and explained how he guided them to the care of two sailors from the warship *Niobe*. The injured Wells was himself rescued by a passing salvage wagon from the Union Protection Company. The company also found fireman Leahy, alive but unconscious, and seriously injured. Both men were taken to Camp Hill Hospital where Leahy died twenty-five days later. Wells spent the next two days on the hospital floor waiting for a bed. He was released from hospital five months later. After the *Patricia* was restored, Wells was presented with the remaining half of the steering wheel. He died of old age in 1971.

As time moved on, and the worst of the Halifax Explosion was behind the city, the fire department started to focus on more day-to-day matters, such as the issue of ringing bells at the city's fire stations.

At a fire wards' meeting on June 13, 1919, concerns were raised about the status of the four stations equipped with bells.

Grand Parade Engine House in Halifax, 1853

The fire chief, John W. Church, was of the opinion that the bell-ringing should stop. It "demoralizes business," he said, adding that ringing bells also created another problem by drawing crowds to the firefighters' route and interfering with their work. Now that the city had paid staff and didn't have to rely solely on volunteers who needed the bells for communicating fires, the bells were unnecessary. The Morris Street station bell, in particular, was reported to be an added discomfort to the patients at the nearby Victoria General Hospital. Despite these arguments, no decision was made at the meeting on whether or not to get rid of the bells.

Perhaps this stalling was due to the important consideration of keeping fire spectators happy, as a fire scene was something many Haligonians would do almost anything not to miss. At a later meeting in June 1919, the fire chief said that he would speak with the local phone company to come up with a special arrangement whereby people interested in viewing a fire could obtain the necessary information from the operator.

THE HALIFAX AND DARTMOUTH Fire Departments relied on each other for help whenever disaster struck. In September 1921, the city, still mindful of the 1917 explosion, braced itself for a similar situation when several early morning explosions rocked the area around the Imperial Oil refinery in Dartmouth. A large oil tanker called the *Victolite* from Yarmouth was unloading a cargo of four million gallons of crude oil at the wharf when the first tank exploded. The explosion caused flaming oil to rise into the sky, lighting the sky for kilometres around. A thick pall of black smoke spread towards Halifax and Dartmouth, leaving residents fearful that they would be the victims of another man-made explosion. Fire spread quickly but was eventually brought under control without any lives being lost.

Another example of the collaboration between the Halifax and Dartmouth departments occurred around the time of the Imperial Oil explosion on February 28, 1929, when a fire broke out at the County Home buildings in Cole Harbour—both fire departments responded to the call. The fire swept through the men's living quarters, the hospital, dining room, and the steam heating plant.

Mrs. Conrod, the home's matron, along with her staff, is reported to have helped residents from the burning building before calling the fire department. Firefighters rushed to the scene along snow-covered roads only to be met with a lack of water and an inability to operate their steam and motorized pumpers. Despite these setbacks, they were able to contain the fire. Wrapped in blankets, the home's residents were transported by sleigh and later taken to the City Home in Halifax. The fire left one hundred residents homeless.

Dartmouth Fire Department float during wartime, c. 1918

Fire Marshal J. A. Rudland, in a report on March 4, 1929, determined that the fire was likely caused by a burning tobacco pipe or a lighted match dropped through a hole in the wall. Considered a fire liability, the fire marshal strongly recommended that the building no longer be used. An agreement was reached with the City of Halifax to house the County Home patients for the next five years, at a rate of $3.50 a week per patient.

AS THE 1920s drew to an end, so did the era of the horse-drawn apparatus. In 1929, they were taken out of service in Halifax. While the horses had been of great help to the department, they also had their downside. A few years earlier, the city's health board had raised concerns about horses and firefighters living too close together. In 1921, the board called for better ventilation in some fire stations where the horses' stalls were considered too close to the men's living quarters and the "odour from the stalls" was seeping into the sleeping quarters. Nonetheless, when the animals officially left the city's fire halls, the firefighters bid them a fond farewell. This poem was reported to have appeared in a local newspaper at the time:

A horse-drawn water cart is filled from a fire hydrant at the corner of Cornwallis and Brunswick Streets, c. 1916.

The Fire Horses Farewell
We played the game and played it
Square, at every call of the gong;
We gave our speed, upheld our
Breed, now our life's not worth a song;
Our time is past, the die is cast;
Perhaps it's just as well.
So to our Halifax friends and the Fire Brigade
   we neigh our last farewell.

Of course, relying on motorized vehicles brought a new set of problems. Fielding complaints about speeding fire trucks racing through the city streets became a concern for Chief Churchill. The chief reported that he had warned his drivers not to exceed thirty-two kilometres per hour as this was the equipment's maximum speed. Nevertheless, a few years later, the chief found himself dealing with a motor vehicle accident that had more to do with carelessness than speed. On July 26, 1923, at 6:10 P.M. while responding to Box 38 near the corner of Prince and Barrington Streets, the No. 4 ladder truck was turning down Prince Street when Chief Churchill's car, driven by Driver McIsaac, collided with nine metres of ladder that was projecting from a passing truck. The accident resulted in two metres of broken-off ladder and damage to the side and back curtains of the chief's car. Damage was estimated at fifteen dollars to the chief's car and forty dollars to the ladder truck. In a report presented to the fire wards committee, the best explanation for the cause of the accident was that the driver of the car applied his foot to the accelerator when he meant to apply the brake; the brake and accelerator on the car were quite close together.

IN THE MIDST of the Great Depression, residents of Sheet Harbour, a village located more than one hundred kilometres east of Dartmouth, decided it was time to start their own fire department; hence, the Sheet Harbour Volunteer Fire Department was born in 1936. It was the first of dozens of small rural stations that would spring up in the coming years in what is now the Halifax Regional Municipality.

Three years behind Sheet Harbour, residents of the small village of Bedford followed suit. Construction of the community's first fire station on the corner of Rutledge and Borden Streets, in the central part of the village, started in 1939. The lot was purchased for $150, and on it a red-brick, block building was built, which consisted of two bays and a single door facing Borden Street. The first meeting was held in the station in October 1939; at that time, Elijah Heffler was elected fire chief, a role he retained until 1943. The Rutledge Street Station still stands today; however, it is now privately owned. As a sign of respect for the building's heritage, the owners left a small statue of a Dalmatian dog and a fire hydrant on the outside wall.

Yet long before the first fire station opened, Bedford community members organized to help each other when fire struck. As early as 1922, a shed in the area was reported to have twenty-five red buckets, two wooden ladders, and a fire axe hanging on the wall.

The department bought its first two pieces of firefighting equipment in 1939—a 1923 White–LaFrance hose and ladder truck and a trailer pump. Because the hose and ladder truck, with its crank starter and open cab, proved to be difficult to handle on Bedford's hills, it was replaced in 1943 with a 1938 Chevrolet hose and ladder truck. It wasn't until 1953 that Bedford bought its first new fire truck, a GMC pumper/tanker. A second one was purchased in 1955. A few years later, volunteer firefighters bought a third one and even built their own tanker truck. The four pieces of equipment called for a new fire station which officially opened in 1962 on the Bedford Highway.

Bedford Fire Department chief Wilfred Greenman and Ralph Heffler stand beside the department's first new GMC pumper, 1953

THE BEDFORD department's first fire phone was located in the home of Fred Emmerson opposite the original fire hall. In the early 1950s, phones and siren buttons were put in the homes of Wilfred Greenman, Angus Mitchell and Irvine Boutilier, expanding the system. An alarm base was also installed in the fire station which could alert the nearby fire districts. The siren alarm system remained in effect until 1967 when instant alert receivers were bought. In 1979, these were replaced by a pager and dispatch system.

On October 5, 1954, a year after this photo was taken, Bedford firefighters were called out to a fire at the Bedford Theatre on the Bedford Highway, a disaster that gutted the popular movie house. Ten years later, the department responded to probably its biggest fire at the Loyola Hall of Saint Ignatius Church. The fire caused $250,000 in damages, destroying the hall, glebe, and a convenience store next door.

As residents of a bedroom community to nearby Halifax, most of Bedford's volunteer firefighters worked outside the town and weren't available to respond to calls during the day. With the community expanding and the number of calls increasing, the volunteer force started to feel the strain. It became increasingly evident that during daytime hours, career firefighters were needed to run the station. On June 5, 1975, the Bedford Service Commission approved the hiring of five career firefighters. They were hired from within the volunteer ranks and worked during the day on weekdays. In March of 1998, two years after Bedford amalgamated to become part of the Halifax Regional Municipality, career firefighters started twenty-four-hour shifts at the station, and the volunteer firefighters were relegated to be back-up for the firefighters. The department shrank from sixty volunteers to thirty.

The 1980s saw several changes within the department—medical emergency calls increased, and the first female firefighter joined the ranks. Due to the increase in medical assistance calls, Bedford firefighters raised money to buy a new defibrillator in 1992. They were trained on the equipment and some even became trained in Advanced Cardiac Life Support, which allowed them to administer drugs, start IVs and perform other medical procedures. Bedford was known for its proactive approach to emergency medical services in the fire service.

Bedford Fire Chief Wilfred Greenman and firefighter Ted Corbin test the capability of their new pumper in 1953

Queen Hotel fire in Halifax, March 2, 1939

MONTHS BEFORE Hitler invaded Poland on September 1, 1939, a fire destroyed one of the city's big, old hotels on March 2, 1939. The Queen Hotel fire on Hollis Street proved to be the greatest blaze Halifax had seen since the 1917 explosion and one of the country's worst hotel fires, leaving at least twenty-eight people dead and several others injured. Built in sections between 1849 and 1908, the main body of the hotel had five storeys, largely brick with wooden interiors and unprotected doorways—in effect, one big fire area. To make matters worse, elevator and stair shafts weren't cut off from smoke or the spread of fire; there were no sprinklers, alarm systems or watermain service, and the outside fire escapes were inaccessible. The conditions were ripe for disaster.

Shortly after 6 A.M., the night clerk discovered smoke coming from the hotel's furnace room. The clerk called the fire department and set out to try to alert the hotel guests. With no alarm system he had to call each guest room on an old-fashioned switchboard. There were seventy-eight registered guests and thirty hotel employees in the building at the time. However, he only had time to make a few calls before smoke forced him to abandon his desk. Within minutes, flames engulfed the five-storey structure.

When the fire department arrived from the Bedford Row station only a block away, people had started jumping from windows. A desperate father was able to get his two children out through a window and into a life net nine metres below before following them to safety. One fireman climbed through one of the hotel's upper-storey windows to see the silhouette of a woman kneeling in prayer surrounded by advancing flames. The heat of the fire forced him outside before he could save her. In one instance, two firemen raced up a ladder, ducking shooting flames to rescue a woman who only moments before had tried to reach a ladder, missed, and was left clinging by her fingertips to a narrow ledge. The firemen had just enough time to slip a rope under her arms and lower her to safety before she passed out.

Despite the huge loss of life, several hotel guests were saved by either firemen or the fire life nets. However, a maze of overhead wires hampered work with the one available aerial ladder. The fire destroyed the hotel and two adjoining buildings, causing $800,000 worth of damage.

"Hollis Street looked as if it had been ripped by a tornado," wrote the *Halifax Herald*. "Debris littered the street and all lanes of traffic were blocked by fire apparatus.... The skeleton walls of the once-famous hotel were reminiscent of scenes in battle-scarred France."

The *Halifax Daily Star* ran the photo on the previous page on March 3, 1939, with the following caption:

"The [above] picture presents a general view of what is left of the Queen Hotel on Hollis Street, which was leveled by fire yesterday morning.... Note where the hotel wall has collapsed, barely missing two automobiles that never will be claimed. Their owners are among the missing. Great difficulty was met by firemen attempting to raise the giant aerial tower and ladder in the foreground."

Following the fire, Justice M. B. Archibald of the Nova Scotia Supreme Court was appointed to investigate not only the fire but how to better safeguard human lives and prevent another such fire. Blame for the fire fell on the shoulders of hotel owner John Simon, as it was discovered that fire-safety features were non-existent in the hotel, and staff were not trained in fire procedures. Justice Archibald's report also noted problems with the

[FACING PAGE] Remains of the Queen Hotel after the devastating fire of 1939

fire department's apparatus:"There was a scarcity of ladders of sufficient length to reach the upper floors of the building....[The] aerial ladder is twenty-years-old, and from the evidence of the firemen and the chief of the department, it is apparent that the ladder is unsafe...

"The presence of overhead live wires, combined with the failure to provide the fire department with more efficient, up-to-date and speedy wire cutting apparatus greatly hampered the firefighters during this fire, and interfered with the work of [rescuing] the occupants of the hotel."

His report criticized the Fire Marshal's Office, accusing it of failing to enforce existing safety regulations and recommended that the office be overhauled to have better accountability and increased enforcement powers. The cause of the fire was never determined.

Training during World War Two takes place on the Halifax's North West Arm.

DURING WORLD WAR TWO, every aspect of life in Halifax and its surrounding communities was affected. Being the home of Canada's main naval port, Halifax saw more than 150 troop convoys embarked from its shores. After Canada declared war on Germany on September 10, 1939, Halifax became a strategic centre for both the British and Canadian navies. Due to the overcrowding of the city during the war, the municipal water supply, along with other services, became overtaxed, subsequently creating difficulties for local firefighters.

During the war, a call went out to Canadian firefighters to assist the British Fire Service during the devastating bombing raids over England. The call appealed to William Palmer Inglis, a hoseman with the Halifax Fire Department. On August 15, 1942, Inglis, along with hoseman Harry Curran, Captain Joseph Harber and fire alarm boss Ernie Peek, all from the Halifax Fire Department, joined the Corps of Canadian Firefighters.

WARTIME Halifax Harbour was patrolled by the fire boats *James Battle* and *Rouille*. *Rouille* was an eighty-five-ton steam tug that had chemical foam onboard, as well as a fire pump with a capacity of two thousand gallons/minute. It was operated by the National Harbours Board, and its firefighting crew was under the direction of Captain Donald Preston of the Halifax Fire Department. *Rouille* went into service in July 1941, and served until August 1945. A second fireboat was added to serve the harbour in August 1943 when the *James Battle*, a retired vessel from the Detroit Fire Department, was refitted for service. It had six pumps and an output of seven thousand gallons/minute.

Halifax fire boat *Rouille*, c.1941

The three most potentially hazardous events involving the *Rouille* and the *James Battle* during the war years were the scuttling of the burning British munitions ship *Trongate* in April 1942, the fire aboard the American munitions ship *Volunteer* in November 1943, and the Bedford Magazine Explosion in July 1945.

On the night of April 9, 1942, Halifax was faced with the prospect of another deadly explosion in its harbour. Off the northeast shore of Georges Island, the *Trongate*, carrying three thousand drums of highly flammable toluol and small-arms ammunition, caught fire. Military and civilian firefighters responded, but failed to extinguish the fire. Since the vessel's boilers had been previously shut down, it could not be moved to a safe location. In the early morning of April 10, the Canadian minesweeper *Chedabucto*, with the help of a US warship, pumped non-explosive shells into both sides of the vessel along its waterline. This sunk the ship, preventing it from exploding and causing damage to Halifax and Dartmouth. For a brief time the residents of Halifax and Dartmouth thought they were under attack. The imprint of the ship— known as the "*Trongate* Depression"—is still clearly evident on the harbour floor, near Georges Island.

On November 3, 1943, fire broke out onboard the American *Volunteer*, threatening five hundred tonnes of light ammunition, eighteen hundred tonnes of heavy howitzer ammunition, dynamite, and two thousand drums of combustible magnesium. Fortunately, the *Volunteer* was towed to McNab's Island, beached, and extinguished before a disaster occurred.

Today the Halifax Regional Fire and Emergency Service has a harbour rescue boat equipped to fight small fires. In the case of a larger fire, it would turn to a Department of National Defence boat.

Air Raid Precautions (ARP) in Halifax

THE HALIFAX CIVIL EMERGENCY CORPS, directed by Major Osborne Cromwell, was the city's civilian defence corps during World War Two. The corps was made up of several divisions that could quickly mobilize to combat air raids and fires, perform rescues, administer first-aid and distribute water. By the end of 1943, six thousand members had enlisted. Worried about incendiary bombs, the unit established an Air Raid Precautions (ARP) network, modeled after a similar system in London, England. In 1942, the corps ordered every household in Halifax to stockpile bags of sand to fight fires.

Halifax Fire Chief John Churchill was in charge of running not only a city wartime fire department but also co-ordinating the ARP auxiliary firefighters. By July, 1942, Halifax ARP had been issued twenty-two 150 gallon/minute pumpers, but it lacked tools such as axes, ladders, and hydrant wrenches. ARP turned to the public for help, and donations flooded in. All federal assistance to the ARP ceased after March 31, 1945, but the thousands of members in Halifax remained on alert until the end of the war. They provided valuable assistance fighting fires, such as the Bedford Magazine Explosion of 1945.

Rather than disband after the war, some ARP units remained together to form the beginnings of a peacetime fire department. The Fairview Fire Protection Association was one such former ARP unit. With just a few tools and some fundraising, they built a new fifteen hundred dollar fire hall complete with the essentials—a siren, a small fire truck, a hose, and a pump.

Air Raid Precautions (ARP) Platoon 7 in Halifax's North End, July 8, 1943

Fairview Fire Protective Association, c. 1946.

"With a force of twenty-seven trained volunteers Fairview residents may well feel that they have a fine, firefighting force, which, will, in due time, with public support making necessary expansion possible, be able to cope with any emergency," the *Halifax Mail* reported on May 31, 1946.

Dartmouth ferry *Governor Cornwallis* on fire, Dec. 23, 1944

At 4:05 P.M., two days before Christmas in 1944, the Dartmouth ferry *Governor Cornwallis* left the Halifax dock with about twenty motor vehicles, and between three hundred and four hundred passengers. Almost immediately, the engine-room crew discovered a fire in the ceiling of the engine room. Rather than raise a general alarm, the two men fought the fire with extinguishers. The boat, which was the ferry commission's first ferry with almost total wheelhouse control, continued on its journey; the captain remained unaware of the events below deck. Subsequently, all vehicles and passengers had disembarked in Dartmouth before it was widely known there was fire onboard.

The fire had grown out of control when Dartmouth firefighters arrived on the scene a few minutes after the boat docked. The engine-room crew was ordered to leave while tugs and fire boats raced to the ferry. Consumed by fire, it was towed to Georges Island and beached.

Rumours of sabotage swirled, but an investigation by the provincial fire marshal on January 12, 1945, concluded that the cause was due to poor installation of the heating furnace's smoke pipe. The fire marshal also gave special mention to the engine-room crew, engineer Carmichael and oiler Horobin: "The conduct of these men is brought to the attention of the Minister of Labour because it was outstanding in character. These two men, without regard for their own personal safety, held this fire in check until and after the ship was docked. In their hands rested the safety of the passengers carried on that voyage. By their efforts they did control this fire and that action probably resulted in the saving of hundreds of lives. They did not leave their posts until ordered to do so."

THE YEAR AFTER the *Cornwallis* sank in 1944, the harbour was the site of another major event. Following VE Day on May 8, 1945, the push was on to decommission the country's war ships upon their return home. Many of them arrived in Halifax ready to unload their unused ammunition of shells, torpedoes, and pyrotechnics at the Bedford magazine, on the eastern shore of Bedford Basin. After the 1917 deadly explosion, when the munitions magazine at the Wellington Barracks threatened to explode, the magazine had been closed and one in less densely populated Bedford opened.

Between May 1 and July 18, close to eighty-three ships reportedly dumped their loads of ammunition at the Bedford depot. The magazine became so crowded that some ammunition had to be stored outside, near a jetty. On a hot July 18, 1945, at about supper time, a pyrotechnic rocket is reported to have ignited, setting off fires and a series of small explosions on the south jetty. At the time, the jetty had a barge tied alongside it.

"A bunch of the guys were out back of the Provost Corps barracks on Cogswell Street, playing horseshoes, when there was this terribly loud explosion around 6:30 P.M. At the same time we could see this large mushroom-like cloud," Russell Harkness of Amherst told the *Chronicle Herald* in January, 2000. "It was a very dark, black cloud with a tail. It drifted high into the sky and stayed there a long time. We all stood there for a moment wondering what was happening."

Bedford Magazine Explosion, July 18, 1945

Harkness, then twenty-five years old and a member of the Canadian Provost Corps, soon learned the magazine had caught fire and was exploding. The north end of the city shook from the explosion and a big thunderclap reverberated across Halifax and Dartmouth. Ignited ammunition unleashed a string of fireworks, sending a huge cloud of dark smoke into the sky. The fireboat *Rouille* sped to the scene. The *James Battle* also appeared but was ordered to stand off at a safe distance. Naval firefighters, under the guidance of the ammunition foremen, worked to control the fires that were quickly spreading in the dry grass and bush surrounding the open piles of ammunition.

The force of the explosion at the magazine knocked out the water main and damaged its supply tank. Eventually the breaks were fixed, and *Rouille* came in to pump water into the remainder of the hydrant system. At 9:00 P.M., the navy ordered an evacuation of the city's north end; eventually, most residents along the Northwest Arm also evacuated, causing a traffic jam. Many went to Citadel Hill, where they could watch the spectacle.

"Bullets were flying into the sky for much of the night. A lot of them were tracer bullets, and as a result you could see them clearly and they were going everywhere. At times it was just like fireworks, very pretty," Harkness told the *Chronicle Herald*. He spent much of the night driving evacuees to the Halifax Commons, which had become a tent city.

The largest explosion came at 4:00 A.M. when a concentration of more than 360 depth charges and bombs went up, leaving a huge crater. The shock-wave travelled across the water and ricocheted through the streets of downtown Halifax, breaking windows as it went. Though bangs and whizzes continued for several days, the worst was over.

Newspaper accounts of the explosion reported that one person died—Henry Craig, a navy patrolman who had been on night watch. On duty at the jetty when fire started on a vessel, he sounded the alarm, and then rushed to fight the fire. He was killed by the first major blast.

Military and civilian firefighters fought the blaze, while dodging bullets and explosions, for almost twenty-four hours. Their heroic work was later praised by officials. One-third of the magazine area was burned, but damage in the city was limited. The day after the explosion, the fifteen thousand evacuees were allowed to return home, where many found shattered windows and cracked plaster.

Explosion of munitions at the Bedford Magazine, July 18–19, 1945

Lakeside Volunteer Fire Department

OUTSIDE THE CORE of Halifax, the community of Lakeside quietly opened its own fire station in 1948. It wasn't the beginning of a fire service in the area; four years earlier, five men in the nearby village of Greenhead had decided to take fire protection into their own hands. They bought a pump and mounted it on a trailer which they would tow behind a truck when racing to a fire. This service was in operation until 1950.

When the Lakeside Station opened in June, 1948, to serve the bedroom communities of Beechville, Lakeside, and Timberlea, Jack Rutt was hired as the first career firefighter. On call twenty-four hours a day, he was aided by twenty-seven volunteers. When a fire call came in, Rutt was called at his home, which was conveniently located next to the station. He would then run next door and ring the alarm, mounted on the station's roof. The site of the fire was marked on a blackboard and as the volunteers arrived at the station they would check the board to find out where to go.

As the trio of communities grew, more equipment was added. In June, 1983, a new fire station was officially opened at 26 Myra Road, in Timberlea, on land donated by the Lakeside Legion. The fire service had to keep up with the fast growth in their communities. In 1991, the population in the area was 6,070, but by 1996, it had jumped to 9,894. By 1998, the department had four career firefighters and about thirty volunteers.

On a sunny Sunday afternoon in July 1996, an explosion shattered Lakeside residents' quiet lives. Leaking propane in the basement of a house ignited and levelled the three-storey building. Thirty other homes were damaged in the explosion. Miraculously, no one was injured.

Captain Ray Cormier, Joe Brigley, Clifford Snair, and David Joy attend junior firefighters' training at the Black Point and District Volunteer Fire Department, 1959

BLACK POINT'S fire protection services can be traced back to November 14, 1949, when the community got a new 1949 GMC Bickle Seagrave fire engine. Without a fire station to house the truck, it was kept in Graham Dauphinee's barn in Boutiliers Point and later moved into the basement of the Bizzee Center Restaurant in Black Point.

The first fire call recorded by the not-yet formally organized department was on February 22, 1950. It was a chimney fire at the home of Loftus Mason. During that year, residents in the Black Point area responded to thirteen calls. It wasn't until 1956 that an official fire department was formed. In March of that year, local mill operator Harold Whittier received training on how to use the fire engine and was hired as an on-call driver. With a driver in place, a meeting was held at the Boutiliers Point School on June 26, 1956, to form a local volunteer department. Wilfred Snooks became the first fire chief. He served for just one year.

Focus now turned to building a fire station. A two-bay station was built onto the former Black Point School and was ready for use in March, 1957. The older part of the building became a hall for meetings, training, and fund-raising. With a hall up and running, the department put more energy into training. By 1958, it became known as a provincial leader in training young people through its junior firefighters' program. The youth program is still running today.

In the late 1970s, the department responded to an increasing number of calls by buying a former Imperial Oil service station and opening a sub-fire station at the Head of St. Margaret's Bay.

From its humble beginnings as a one-engine service answering about a dozen calls a year, the department has grown to one that answers more than two hundreds calls per year, with four paid staff and a team of volunteers.

Chief John Churchill

HALIFAX FIRE CHIEF John Churchill, who had become chief of the department on December 12, 1917, following the death of Chief Edward Condon in the Halifax Explosion, died on September 26, 1945, after a brief illness. On January 1, 1946, Fred MacGillvray took over as chief of the department.

Credited for never having missed a large fire during his career, Chief Churchill brought in several substantial changes during his reign. He saw the Halifax Fire Department motorized, brought in a fully-paid fire service, and started a fire prevention division. Through the division, public buildings, starting with schools, were equipped with sprinklers and alarm systems.

Chapter 2

Fresh Starts:
Development of Rural and Urban Fire Departments

When 1950 rolled in, Halifax was still recovering from celebrations the year before that had marked the two hundredth anniversary of European settlement. While the city pulled out all the stops to celebrate the anniversary, it couldn't hide the fact that it was still feeling the scars of the long, hard depression and World War Two. In the decade following the war, the pace of change was slow in Halifax and wouldn't really pick up until the 1960s. City council had adopted a master plan for postwar construction in 1946, and in the following years, the rebuilding process began.

Fire Prevention Float in Halifax following the war

Part of what had to be rebuilt was the city's infrastructure. It wasn't just roads and housing that needed help; its crumbling nineteenth-century waterworks were also badly in need of repair. Half the water being pumped through the city's water supply system was reported to be leaking into the ground, proving problematic for firefighters relying on the supply to do their work.

Urban renewal and suburbanization in the coming years would change the look of the city and create an increased demand for adequate fire services in the growing areas close to Halifax. The next two decades would see several small, rural departments spring up—all fuelled by the desire of community members to protect their neighbours in times of need.

Around this time, the fire department was also turning its focus from not just the primary concern of fighting fires, but to preventing them from occurring in the first place. The idea that it was better to prevent fires than respond to them when they occurred was catching on not just in Halifax but also across the country. From the earliest times, fire-prevention bylaws were written locally and fell under the responsibility of municipal governments. After the middle of the nineteenth century, better building and fire prevention codes were developed by the National Board of Fire Underwriters in the United States. These were guided by engineers in consultation with fire chiefs and building experts. While Nova Scotia's first fire prevention act was proclaimed in 1919, the same year as the first national fire prevention day was held, it wasn't until after World War Two that fire codes were introduced and real progress was made in the area.

In Canada, a federal fire research section was established in 1951 to develop and implement a National Building Code and the National Fire Code. These policies would become widely adopted by cities and towns across the country.

Fire destroys Motherhouse, Jan. 31, 1951

FIRE COULD NOT be prevented on a cold January night in 1951. Flames caused more than four million dollars in damages, destroying Mount Saint Vincent College's (now Mount Saint Vincent University's) main complex, known as the Motherhouse. The temperature was minus twenty-five degrees Celsius when more than four hundred Sisters of Charity, students, and novices had to leave their dormitories. Many of them wouldn't leave the burning building until they were cloaked in their habits and veils. Once properly dressed, they evacuated the building in an orderly fashion through smoke-filled corridors. They were billeted at St. Joseph's Orphanage on Quinpool Road. Classes resumed on campus in the fall of 1951, but the Motherhouse didn't reopen until 1958.

When firemen arrived on the scene, the house was already beyond saving. Only a station of the cross in the courtyard and a hand-painted ornament survived the fire.

A firefighter poses with Bedford Row Station's Dalmation, c. 1950

WHEN WEARY FIREFIGHTERS return to their fire halls after hours spent battling blazes, the traditional image is that they will be greeted by a friendly Dalmatian, which has long been linked to fire-station lore. Considered to have the necessary stamina to endure the rigours of firefighting, as well as an innate ability to run among horses' hoofs without being trampled, Dalmations were the companion of choice for firefighters in the days of horse-drawn apparatus. They were also said to form strong bonds with fire horses and were a calming influence during critical times. In the early 1900s, however, the Halifax Fire Department's mascots were not Dalmatians but instead two Jack Russell terriers, Nix and Girlie, who appear in several photos with then-Chief Patrick Broderick.

Dalmatians do not appear to have played a prominent role in the fire department's history. However, in the 1950s, the wealthy Oland family of Halifax donated a Dalmatian to each Halifax fire station. The dogs apparently developed sore feet and cracked pads from the station's concrete floors, and had to be given away.

In recent years, Minnie the cat was the mascot of the West Street Station in Halifax. According to a 1997 story in the *Chronicle Herald*, she had wandered into the station eighteen years earlier.

Fresh Starts · Halifax Regional Fire & Emergency

The crew of 9 Quint, 'C' Platoon, pose in front of their station in Lower Sackville on Sept. 23, 2005. From left to right are Firefighters Wilfred DeBay and Earle Slauenwhite, and Capt. Gregory Hebb.

ON DECEMBER 6, 1954, a group of concerned citizens gathered at a local hall in Sackville to discuss the formation of a fire department, expressing concerns about the length of time it took for fire trucks to respond from nearby Bedford. After the meeting, the residents voted in favour of a local fire organization; by the spring of 1955, the department was formed.

The first fire truck of the station was housed in a small shed and was later moved to a barn on the Lynwood Farm on the Old Sackville Road. Firefighters soon constructed a new two-bay fire station at the corner of Beaverbank Road and Gloria Avenue. The station was home to a pumper and a tanker truck, one of the first in Halifax County.

In the 1960s, a major building boom took place in Sackville, and by the 1970s, the population had tripled in size. Growing in order to keep pace with the population, the station underwent an expansion, adding more space for equipment, as well as a meeting area and recreation hall. The first career firefighter was also hired at this time. By the early 1970s, however, the department was flooded with volunteers, numbers of which peaked at eighty. In 1975, the station underwent a third expansion to add sleeping quarters and offices. Firefighters also took on the added responsibility of manning a newly-established ambulance service. With housing construction continuing in the area, a second station was built at the corner of Highway 1 and Old Patton Road.

The Sackville Fire Department continued to grow throughout the 1980s and 1990s. It took on water rescues, created fire prevention and training divisions, and formed the Sackville Volunteer Firefighters Organization. In the early 1990s, two new fire stations were opened on Metropolitan Avenue and Sackville Drive, and the Beaverbank Road Station was closed.

In 2005, the department marks its fiftieth anniversary.

IN WAVERLEY, residents were also starting to contemplate better fire protection for their community. On May 2, 1955, A. T. Milligan, Harold Gladwin and Lou Meagher, three long-time residents of the village, called a public meeting. Following a unanimous vote, it was decided that fire protection for the area was absolutely necessary and that immediate steps should be taken to organize a volunteer department.

Waverley Volunteer Fire Department

It was decided that every householder should pay five dollars to raise funds to start the fire department. In the following months, nine hundred dollars were collected from the community, enabling the department to buy a pumper from the Bedford Fire Department for $300, as well as a jeep to haul hoses and nozzles, for $150.

At a meeting on February 26, 1956, Lou Meagher was chosen as the department's first chief and Wes Hilchie as his deputy. Four years later, a station, consisting of three small bays and a radio room, was built.

Fire station in the community of Goffs

THE COMMUNITY OF GOFFS, located about twenty-eight kilometres northeast of Halifax at the beginning of the Old Guysborough Road, also built a fire station which is now considered a sub-station of the Waverley Volunteer Fire Department. In 1980, three women joined the Waverley department. Six years later, the department hired two career firefighters during the day to assist while the volunteers were at work. The department now has twenty-five volunteer members.

Two of the Waverley department's more famous members are Don and Dave Carroll of the band *Sons of Maxwell,* who signed up as volunteers with the department in September 2004. Originally from Timmins, Ontario, the brothers moved to the Halifax area in the mid-1990s. They decided to join the department as a way of becoming a part of the community.

As children, the Carroll brothers became interested in firefighting when they tagged along with their grandmother while she cleaned the local fire station. Since joining the service, they have seen their fair share of action, including a tragic plane crash at the nearby Halifax International Airport in October 2004, in which all seven crew members onboard the plane died. Not only were the brothers on the scene, but they made it onto the national news. As new, not yet trained firefighters, they wore black helmets. However, some reporters thought their differently coloured helmets meant they were in charge and started barraging them with questions.

THE FISHING VILLAGE OF TERENCE BAY was the first community in the Prospect Road area to get a fire hall. After community members chipped in, the hall was built in 1956. Before they had their own space, they operated out of a garage with a hand-built fire truck. Today the station is non-operational because of a lack of volunteers, but the department's current chief, Barb Sawatsky, says they are working hard to recruit and train new members.

Hatchet Lake Fire Department on the Prospect Road was the next community to build a fire hall. Using recycled materials from a building torn down in Halifax, volunteers completed the hall in 1966. For years the two departments of Terence Bay and Hatchet Lake operated independently of each other. The older members still tell stories of racing each other to calls on the Prospect Road to see who could arrive first on the scene. This eagerness to get to a fire as quickly as possible has not faded over the years: some volunteer firefighters will take whatever mode of transportation they can, including their all-terrain vehicles.

As the population in the area grew and more people started working further away from their communities, in 1989, the department saw the need to hire four career firefighters to oversee the two stations. Two years later, under the direction of then Terence Bay Chief Doug Avery and former Hatchet Lake Chief Bob Barr, the two departments were amalgamated and a third fire station in Shad Bay was opened, headed up by Avery. The new combined department had four career firefighters and more than seventy-five volunteers.

Water remains a key issue for firefighters, not just in the Prospect area but in all rural areas. With no municipal water and sewage services, there are no fire hydrants in such areas, meaning firefighters have to rely on other methods. One solution is to put in dry hydrants which can draw water from a lake or pond. Due to a lack of water in these areas, tanker trucks are used to start firefighters off in their attack of a fire until other water sources arrive.

Another challenge for rural firefighters is managing the hordes of spectators at a scene. "Usually you have the whole community following you to a fire," says Christine Scott, a Prospect Road firefighter and safety officer. "Problems can arise when people park their cars along the road and block fire trucks."

Today the Prospect department responds to calls from Exhibition Park on the Prospect Road to the community of West Dover and all seventeen communities in between, a huge area stretching seven hundred square kilometres. In 2004, the department responded to almost three hundred calls.

Barb Sawatsky, chief of the Prospect Road and District Volunteer Fire Department

Wellington and Fletcher's Lake Volunteer Fire Department, 1970

THE RESIDENTS OF WELLINGTON, a small community of about 150 families, about twenty-nine kilometres from Halifax, started a fire department in the mid-1950s. The project was spearheaded by Men's Community Cribbage Club, which was prompted to make the move after a new home burned to the ground due to a lack of firefighting equipment. Through bingo games, donations, and raffle ticket sales, enough money was raised to buy an old fire truck for one hundred dollars.

The department's first fire chief George Roscoe, a blacksmith in the Royal Canadian Navy, told a newspaper reporter in 1957 that the department was built bit by bit. "When we got the truck it wouldn't go, but the boys worked on it, so now it goes like a charm," he said.

In 1956, the department operated out of Roscoe's home. It wasn't until the following year when it got its formal start, operating with twenty-six men and a truck equipped with a five-hundred-gallon tank and pump. The firefighters had to rely on the telephone operator and a member of the department to spread fire alarms. They hoped to have enough money to buy a whistle, which they believed would have the crew "ready to go" more quickly.

As the 1950s drew to a close, the department started to build its own station—after finishing their day jobs, the firefighters would return home in the evenings to work on its construction. Until 1985, two years before its thirtieth anniversary, the department was known as the Wellington Volunteer Fire Department, when Fletcher's Lake was added to the name. In 1987, the year the department marked its thirtieth anniversary, it hired its first female volunteer firefighter. Wendy Stonehouse, a mother of two, told *The Daily News* that she joined because she hoped to help her community. "The men have been really good," she told the paper in September, 1987. "They've gone out of their way to make me feel comfortable. Being the only woman is actually kind of a drawback." She added: "I'm actually surprised I'm enjoying all of it."

Eastern Passage/Cow Bay Volunteer Fire Department, c. 1985

HRM's CURRENT chief director, Michael Eddy, grew up in Eastern Passage and remembers, as a young boy in the late 1950s, sitting in a car with his parents watching as their church burned. The fire left an impression on him, and not long afterwards rumblings started that the community needed its own fire department.

In the fall of 1959, members of various local church groups decided it was time to stop relying on the services of the nearby Shearwater military base (now known as CFB Shearwater) and the Dartmouth Fire Department. A number of meetings were held during the winter of 1959 and 1960, and in the early spring of 1960, a new fire truck, a Chevrolet chassis with American Marsh pump was ordered. They arrived in the late summer of 1960, and a building was rented in what was formerly known as the military's A-23 Training Centre. The first regular meeting of the volunteer fire department was held on December 7, 1960. Headed by Reginald Hunter, the first chief, and Walter Langille, his deputy, they had thirty volunteers. Harry MacDonald, a local resident, agreed to have an extension of the fire hall telephone placed in his home so he could easily answer it and warn firefighters of any fires. In April, 1961, the department's ladies' auxiliary was formed.

Lakeview / Windsor Junction / Fall River Volunteer Fire Department

LIKE MANY COMMUNITIES around Halifax following World War Two, the villages of Lakeview, Windsor Junction, and Fall River witnessed significant growth, increasing the need for reliable fire protection. The old era of bucket brigades was no longer adequate in an area that now contained six hundred homes, several schools and churches, and nearly twenty businesses.

Early in 1961, at a meeting held at the Products Tank Car Shop (Procor), the first official move was made to organize the department. A constitution was drawn up that same year by Earl Hartling, Percy Metzler, Ned Bourgeois, and George Wyatt.

Rodney Smith, the department's first chief, worked with his deputy, Jack Shea, and twenty-five volunteers. In the early days, each member had to pay twenty-five cents a week to belong to the department. A member who missed a regular duty night was punished with tasks such as painting a fire hall room. One of the members' first tasks was the converting of an old International oil truck into a working fire truck—other pieces of equipment were a couple of fire extinguishers and a pump.

In 1964, the department built a fire hall on Fall River Road, in Fall River, on land donated by Percy Metzler. The original fire hall is still used today, but the department's main station is now on Highway 2. Plans are currently underway to build a new building that will house both a fire station and a recreation centre. Construction is expected to begin in the spring of 2006.

George Wyatt drives the Lakeview / Windsor Junction / Fall River Volunteer Fire Department's rescue unit No. 2, a 1955 Fargo, c. 1969

REMEMBERING THE DAYS of the primitive fire phones, Nancy MacDonald, former president of the department's ladies' auxiliary, recalled how the phone would ring in the middle of a cold winter night. She would hop out of bed, pull her snow suit over her nightgown and run next door to the fire hall to blow the siren. Later on, when the siren was moved to her house, her job was made a little easier.

MacDonald laughs when looking back on the department's early days. She likes telling the story of the firefighters who arrived at a fire scene in the middle of the night. They eagerly grabbed one end of the hose off the truck and started hauling it down the hill. Moments later, a few other firefighters, not knowing what their colleagues were doing, grabbed the other end of the hose and proceeded down the hill. "The firemen said, 'You shouldn't tell that story,' but it was so funny," MacDonald said.

The department's ladies' auxiliary was formed in 1964 with Aileen O'Leary acting as the president of the nineteen-member group. Four years later, the women started a successful catering business that helped raise money for the department. MacDonald remembers countless nights of bringing firefighters cool drinks to stave off dehydration, and sandwiches to provide energy for a night-long battle with a stubborn fire.

In 1981, Julie Vials joined the department after firefighters responded to a fire in her Windsor Junction home and became an officer in 1987. At the time, the department believed she was the only female officer in a Canadian fire department.

Hammonds Plains Volunteer Fire Department

DONNIE HAVERSTOCK was fourteen years old when he fought his first fire in Hammonds Plains. Now in his late seventies, Haverstock is perhaps the oldest "honorary" active firefighter in HRM, writer John Giggey asserted in 2004.

Long before there was actually a fire department in his community, Haverstock helped neighbours in need. The first fire he fought was on December 8, 1941, one day after the Japanese attack on Pearl Harbour. A house fire in Pockwock killed two children, and local residents stood helpless without a community fire department to respond to the disaster. "Everybody knew then we had to have a fire department," said Haverstock, who served as a firefighter for fifty years.

During the war, supplies were scarce; it took about three years to get the community of three hundred organized and in possession of some basic equipment, such as a hose and a pump. Not much later, another fire broke out—this time at a lumber mill in Pockwock. Haverstock rushed to the scene to help and found himself working a fire hose. Exhilarated by the taste of firefighting, he joined the department when it was officially opened three years later, in 1944.

In June 2000, a forest fire near the Kingswood subdivision forced about eight hundred residents from their homes. Through a stroke of good luck, Lieutenant Tim Scott of the Hammonds Plains Fire Department was testing the brakes on a newly repaired fire truck on St. George Boulevard when the fire started. He raced to the scene and was the first to arrive. He got two homebuilders working nearby to help him get a hose off the truck while backup travelled to the scene. The fire was reported to have burned more than 120 hectares of land before firefighters got it under control.

Today, a four-bay station, complete with classrooms and administration offices, sits next to the original Hammonds Plains Station, built in 1961. Weekdays, two career firefighters work in the station to support the volunteer force. In the evenings, weekends, and holidays, fire protection for the community's twenty-four thousand residents is provided completely by volunteers.

Upper Musquodoboit Volunteer Fire Department

ON MAY 16, 1967 the first official meeting of the Upper Musquodoboit Fire Department was held at the community hall. They were motivated by a recent death. Months earlier, a local resident was killed when his house burned to the ground and there was no fire department to respond.

With no hall and little equipment, Harry Reynolds, the department's first fire chief, and his team of volunteers decided to initially use community member Donald Hutchinson's garage to store gear during the department's first summer in existence.

The department's first truck was an Emo on loan from Halifax. They eventually got gear, consisting of hats, boots and coats, for the firefighters from the navy surplus store. In July 1967, the department got the go ahead to purchase a pumper truck for $501.95. They bought a 1941 truck that had the capacity to pump 1,000 gallons per minute. It wasn't until 1975 when they would buy their first new truck, which had the capacity to pump 650 gallons per minute.

With a truck they needed a hall. Imperial Oil came to the rescue. The company sold the department a former service station and garage for just $1. After months of renovations, they opened the hall and held their first meeting on Dec. 4, 1967. It wouldn't be until October 1981 that plans for a new fire hall were started. The following year the department bought the land where the current hall stands from Murray Preeper.

When the fire department first started, the volunteers were called to a fire by a siren that was switched on by the local telephone operator. After being alerted by phone about a fire, the operator would flick a switch next to her switch board and activate the siren. That system was in place until 1976 when the phone company installed six fire phones in homes throughout the community. After a fire was called in, the person answering one of the phones simply had to pull and switch and the siren would sound. They used that system until the department bought pagers and firemen were directly alerted about an emergency by their pagers.

Like most volunteer fire departments in the region, fund-raising played a major role in not only the survival but the evolution of the department. The ladies auxiliary, formed in the early 1980s, played a central role in these activities. Through barbeques, 50-mile yard sales and raffles, the department raised money not just to buy new equipment and upkeep the hall, but for the larger community. Some of the funds have gone toward buying an operating light in the local hospital, supporting youth groups and giving out treats to children at Christmas time.

IN THE EARLY 1960s, the Moser River and District Volunteer Fire Department got its start. Ned Lowe, one of the original members, remembers the days when the area had nothing but buckets to battle house fires. In the case of a forest fire, the firefighters would haul out a portable pump and place it in a nearby brook or lake to get water. When the forest fire was extinguished, they used backpacks, which carried about five gallons of water, to put out the smoldering hot spots. Eventually the department built a two-bay fire hall. It is still in use today and now has a community hall attached to it.

Around the same time as residents in Moser River, several inhabitants of Middle Musquodoboit started thinking seriously about starting up a fire service. They organized a meeting of local ratepayers, and after several heated arguments, it was decided that a tax rate of three cents for every assessed one hundred dollars should be levied for the start of the department. It was also decided that this rate could not be increased. Aside from the money raised through taxes, work began on fundraising picnics, dances, and raffles.

Middle Musquodoboit Volunteer Fire Department, 1951

The first regular meeting of the Middle Musquodoboit department was held on May 18, 1961, at the church hall. Chief Charles Milner chaired the meeting and counted thirty-seven members. Even before the first regular meeting was held, the department got a 1956 Chevrolet truck which had a front-mounted pump and a thousand-gallon storage tank; it was kept in a local resident's garage. Building a fire hall was the next concern. Estimates came in at $4,500 for materials. To add income potential to the building, members decided to add a second storey to the hall so they could rent office space to the Department of Agriculture and the Department of Lands and Forests.

The first meeting in the new fire hall was held on January 2, 1963, at which time a fire siren was donated and placed on the hall's roof. Like most rural departments, the early days were marked by financial constraints. Having a phone in the hall meant a $3.35 charge per month—an expense considered too costly. Another major problem was the lack of water supply in the village's western end. To address this issue, a pond was dug at the back of George Fulton's barn.

The Lakeside Volunteer Fire Department fire-prevention float shows that the job of volunteer fire departments was more than simply fighting fires

RESOURCEFULNESS was not only handy when it came to building a station and securing a water supply, but it also played a key role in the department's dispatch system. When the department was first organized, the phone system in the area consisted of party lines and a central switchboard operator. When fire broke out, the operator was contacted; she in turn would activate the siren mounted on the roof of local resident Hugh Kaulbach's barn. Upon hearing the siren, volunteers would call the operator to find out the fire's location. She also had a list of the members' numbers to call. When the telephone company changed over to a dial system in the early 1970s, the department worried, knowing it would lose the twenty-four hour answering service the telephone operators provided. Their concerns were alleviated by the nursing staff at the Musquodoboit Valley Memorial Hospital, which agreed to help out at no cost to the department. Subsequently, the telephone company installed an emergency fire phone at the hospital's twenty-four hour nursing station with an extension to the fire station. They also installed a switch so the nurses could remotely activate the siren on the fire hall's roof.

Local residents were told the number to call in case of an emergency. When a call came in, the nurses would activate the siren and the first member to the hall would call the hospital to get the details. Over the years, the firefighters expressed their appreciation to the nurses by donating equipment such as a wheelchair, patient lift, and video camera. The system, which went into effect on June 22, 1974, stayed in place until the fall of 2000, when the department changed over to the central dispatch system in Bedford.

Tragedy struck the community in November, 1975 when a local doctor was killed in a house fire—the only fire in the department's forty-three year history in which a life was lost. Accordingly, much debate was sparked concerning the firefighters' quality of training and preparedness. As a result, some members of the department attended courses at the Nova Scotia Firefighters School the following year. Today, the department has about two dozen volunteer members.

Oyster Pond and Tangier volunteer fire departments respond to a motor vehicle accident

FOLLOWING THE TREND of small rural volunteer stations springing up in the 1960s, the residents around Oyster Pond on the Eastern Shore started their own fire department. The department, currently headed by Chief Edgar Kerr, covers an area consisting mostly of small fishing communities and single-family dwellings.

Close on the heels of Oyster Pond, the community of Musquodoboit Harbour, located at the mouth of the Musquodoboit River about thirty-two kilometres east of Dartmouth, got its start in 1961. Bill Turner was elected the first fire chief; he and a handful of volunteers built a structure large enough to house an engine. Today, Myles Faulkner is the department's chief, and has about twenty-five volunteers working with him.

EVEN BEFORE the community of Ostrea Lake, situated on the shores of Musquodoboit Harbour about forty kilometres southeast of Dartmouth, had its own fire hall, a group of local residents travelled to the Musquodoboit Harbour Volunteer Fire Department to attend training sessions. As the population grew, the need to establish a fire station became apparent, and with the efforts of local residents, a ten-horse-power portable pump and trailer were purchased and stored in a barn next to the general store. Interest in the project grew, and the Ostrea Lake / Pleasant Point Volunteer Fire Department was established in 1961. An old schoolhouse was bought to serve as the department's headquarters; a fire truck from the Windsor Fire Department was also purchased. This truck was a 1941 International, with an

Fire consumes a home in Chezzetcook, c. 2003

open cap and wooden spoke wheels. Today, Ian Lobban is the department's chief, and he oversees about twenty-five volunteer firefighters.

The Acadian community that surrounds Chezzetcook Inlet, about twenty kilometres from Dartmouth, got its own fire department in 1961. The original station, which has since been renovated three times, was built under Chief René Dubois. The department started out with a dozen volunteers and now has close to thirty. It also has a couple of career firefighters who serve not only the community of Chezzetcook, but the surrounding areas as well. Alan Deschesne is the department's current chief.

In 1962, the fire department for Westphal / Cole Harbour was organized by the District 14 Service Commission, which acted as a governing body for communities in the area. Before the department opened, the area relied on the Dartmouth Fire Department to respond to any fires. The Dartmouth department continued to respond to emergency calls in Cole Harbour, Eastern Passage, and Lawrencetown, even after these communities obtained their own fire services, to provide use of its "Jaws of Life" in motor vehicle accidents.

Initially a fully volunteer station, the Westphal / Cole Harbour department eventually hired full-time paid staff to run the station. Its first building was a two-bay garage on Cole Harbour Road which was replaced by a new station, built on the same road, in 1987.

A ladies' auxiliary for the department was soon formed to help raise money to buy a pumper truck and other equipment. A major fundraiser in the early days, during the 1960s, was hauling water to fill wells during the dry summer months. Former chief Murray Elliott remembers the department's tanker truck, which could hold twelve hundred gallons of water going steady all summer—and even through the winter!—filling local residents' wells. At just ten dollars per fill-up, it was a true deal for community members. By the early 1970s, delivery ceased; not only was it hard on the equipment, there were also worries about the safety of using water from a lake.

"A couple of nights I think I worked all night just hauling water," said Murray Elliott, who started with the department in 1968 as a volunteer. Two years later he became one of the department's paid members and eventually chief in 1979, a position he held until 1996. When Elliott joined the department, situated in what was then mostly a farming community, he was one of about fifteen men. In the early days, the community was so small he knew almost everyone by name. The year before he retired, the department had twenty-eight full-time members, seventy-five volunteers, and more still on waiting lists.

The department's volunteers served Westphal, Cole Harbour, and the Preston / Cherry Brook area, and also made mutual aid calls to assist fire departments in nearby areas. Most calls in the early days were for grass fires on farms and brush fires in the summer months. A structure fire in those days often meant firefighters had to hold their breath for as long as they could while they rushed into a burning house without a breathing apparatus. As Elliott explains, firefighting necessitates getting right into a structure, to find out where exactly the fire is: "You can't think the water is going to find a fire."

Murray Elliott, former chief of the Westphal/Cole Harbour Fire Department

Cole Harbour struck by a tragic house fire, 1989

UNFORTUNATELY FOR FIREFIGHTERS, things don't always go according to plan. On a Sunday morning in April, 1982, the Westphal / Cole Harbour Fire Department answered a fire call at the ruins of an old farmhouse in Westphal. On site, they found a crowd of rowdy partiers who had lit a bonfire in the foundation of the abandoned house. When three firefighters got out of their truck to investigate, someone jumped into it, and proceeded to drive it off the road and into a clump of trees. There were no injuries, but about twenty-five hundred dollars' worth of damage was done to the truck.

In 1989, the department responded to the most tragic fire Elliott had ever seen. The call came in around 7:20 A.M.. When firefighters arrived at 154 Atholea Drive in Cole Harbour, they found the house in flames. Three people—including a mother and her two children— were found unconscious by the back door. The children were revived with oxygen by firefighters; the mother, in full arrest, was revived by CPR. The children's father and a baby girl were found dead in another part of the house. One week after the fire, the mother died due to critical burns from the fire.

Horrific fires such as the one on Atholea Drive made the work of fire prevention all the more pressing. The department raised awareness by going door-to-door throughout the community, one of the fastest growing areas of the province. As well, the department's members tried to reach as many houses as possible to do fire-safety inspections. In 1995, Elliott told the *Chronicle Herald* that in the previous year, fire losses in Westphal / Cole Harbour had reached about ninety-four thousand dollars, of which about forty thousand was arson-related. Clearly, the department's fire-prevention efforts paid off, as this figure was a dramatic reduction from the $674,000 worth of recorded fire losses in 1993.

Beaverbank Kinsac Volunteer Fire Department

IN THE EARLY 1960s, a community meeting was held in a Beaverbank schoolhouse where the provincial fire marshal spoke about the value of having a volunteer fire department in the community. George Hull, who would later serve as the department's chief from 1969 until 1981, and Ken Margeson, who later became a councillor with the Halifax County Council, attended the meeting with interest. At a local ratepayers' meeting on April 15, 1965, the two men pushed the community to create a volunteer fire department. Local need became even more apparent after a number of recent fires in the area. The Sackville Station was about ten kilometres down the road, and the nearby RCAF Fire Station, which had served the Armed Forces base, had closed. Lane MacDonald became the department's first fire chief and he had a team of twenty-seven volunteers plus the original organizing committee.

The community bought a fifteen-hundred-gallon pumper from the Windsor Fire Department and other equipment from the Sackville Department. Financial help came from a tax rate and donations from community members. By 1990, when the department celebrated its twenty-fifth anniversary, it had fifty-six active members and operated five trucks out of two fire halls.

The firefighters' partners became involved, forming the ladies' auxiliary in 1965, which organized dances, fun fairs, and craft sales to raise money. They were also on hand, day or night, to feed tired and hungry firemen.

A huge fire struck the community on May 3, 1978, when flames broke out in a cement plant, located behind Beaverbank Villa. The Sackville Department aided the Beaverbank Department for about three hours until the fire was under control. The diesel-fueled fire burned for about two days before it was successfully extinguished.

Aside from fires, volunteers in the area have been called on over the years to attend to everything from births to searches for missing children. On Canada Day in 1986, the department helped search the local woods after Andy Warburton, a nine-year-old boy visiting from Hamilton, was reported missing. Firefighters, joined by RCMP and several other organizations, spent a wet, cold night searching the woods, but found nothing. The search, which at the time was the province's largest ground search, lasted eight days and was only called off after the boy's body was found. He had died from exposure. The firefighters, along with the entire community of Beaverbank, mourned the loss.

On September 18, 1985, a call came in for assistance at Blaine and Shelly Lively's house in Beaverbank. Prepared to head to the hospital to give birth, Shelly decided to take a ride in the fire department's ambulance when her contractions became just three minutes apart. Riding along in the ambulance, the baby started to crown. Firefighters Laurie Hartlen and Jim Stone, who acted as Lively's attendants, realized that birth was imminent. They were right—at 8:05 P.M., Charmaine Grace Lively was born. Stone later joked that being the catcher of his ball team helped prepare him for the birth. To commemorate the memorable experience, the firefighters hung a picture of Charmaine on the fire hall wall.

Cooks Brook and District Volunteer Fire Department

IN OCTOBER 1963, a number of local grass and chimney fires prompted Herb Jodrey and Vince Dikman to buy a small, portable pump and some hose. They didn't know then but they were creating a fire department that would still be in existence more than forty years later. In 1971, the department, which serves the communities of Cooks Brook, Gays River, and Lake Egmont, bought a milk tanker which served as the first fire truck. Today, the department has eighteen trained volunteers.

THE FIRST RATEPAYERS' MEETING of what would become Canada's first all-black fire department was held in 1964, at a canteen next door to the current fire station on Pockwock Road. The black community, nestled just above Hammonds Plains, became one of the several sites in Nova Scotia where escaped slaves from the United States settled. After the War of 1812, over five hundred black people settled around Hammonds Plains.

The first ratepayers' meeting drew about twenty-five people, a good turnout for a community that at the time was made up of about ninety homes. Interest in starting a department was strong, likely due to several devastating fires during 1963 and 1964. "We had no fire protection in Upper Hammonds Plains then," said founding member Neil Harry Anderson. Nearby Hammonds Plains had a small volunteer department but it couldn't always be relied on. "Sometimes they wouldn't want to come and help, sometimes they would come," he says, adding that the same was true of departments in Bedford and Sackville. "Race played a part at that time," said Wendell David, the department's current fire chief.

Upper Hammonds Plains Volunteer Fire Department Firemen's Memorial Service, October 1975

When residents in Upper Hammonds Plains decided they wanted to organize a department, they faced discrimination from nearby communities who didn't have faith that they could get it going. "They thought we were too poor a community," Anderson said. Despite the naysayers, by the end of 1966, the volunteer department was organized, with forty-eight members. Land on Pockwock Road, which had formerly housed a cooper shop, was donated by a local resident. With donated supplies, volunteer labour, fund-raising, and money from a tax levy, the community built a two-bay station.

The official opening was in 1970 and attended by then-premier, George Smith. "Your men can stand proudly shoulder-to-shoulder with five thousand other firefighters across the length and breadth of this province," Smith told the department, according to an article appearing in the *Mail Star*.

Upper Hammonds Plains Volunteer Fire Department, 1969. Left to right: Alceid Williams, Secretary/Treasurer; George Marsman, Fire Chief; Neil H. Anderson, Liaison Officer; Wesley H. Anderson, Deputy Fire Chief

THE UPPER HAMMONDS PLAINS Department's hall represented nearly four years of volunteer efforts and fund-raising projects. The community observed the opening over a three-day period, complete with a formal dedication and ribbon-cutting, speeches, a parade, and music and dancing.

The Bedford Fire Department donated a thousand-gallon tanker truck, which became the new department's first truck. Other departments donated gear and equipment. With George Marsman heading it, the department started with an annual budget of less than ten thousand dollars.

As a young man in his twenties, Anderson wanted to become a volunteer firefighter to help his community. Over the years, he not only helped to fight fires but also to rebuild four homes ravaged by flames. In the 1970s, Anderson can still remember the confused looks on people's faces when members of the department took part in Natal Day parades in Bedford, Sackville, and Halifax. "Everyone seeing a black fire department riding through," he said. "I guess they were shocked."

Looking back, Anderson remembers one particular fire from the early 1980s. "A house is burning on the hill!" he heard his neighbours calling out. He soon learned it was the home of Roland Allison, fire chief Wendell David's grandfather. When the firefighters reached the house, they found Allison sitting in his chair inside the burning house. "I'll go down with the house," he declared. Ignoring his request, the firefighters carried him out in his chair. Just as they reached the front lawn, the whole house went up in flames. He lost everything in the blaze which had started as a result of a woodstove fire.

Retired Upper Hammonds Plains fire chief Wesley Anderson (left), retired liaison officer Neil Anderson (centre); and Wendell David, current chief of what is now Station 51, stand next to their station's pumper/tanker.

CHIEF WENDELL DAVID has also seen his share of fires, including the February 1961 woodstove fire that left his family without a home. On a cold Sunday morning, a relative living across the street saved the family of eight by racing over with a ladder and helping them to safety.

David, who has been with the department for more than thirty years, became chief in 1992. At that time they scraped by with a budget of twenty thousand dollars. After the amalgamation of HRM, Hammonds Plains and Upper Hammonds Plains fire departments joined forces and now work closely together.

The main problem the department faces now is a lack of volunteers. With only a handful of volunteers to deal with a large number of fires, (in 2004, the Upper Hammonds Plains department responded to about eighty calls) David is actively recruiting. To do so, he is turning to both the black and white communities. While the department has a unique history as an all-black fire department, times have changed. Still, that history can never be taken away. "It makes the community proud," Anderson says of his community's heritage.

In 2001, the department and its women's auxiliary received praise from Chief Director Michael Eddy for their outstanding work at a summer forest-fire at Pockwock Lake. Firefighters spent several days battling smoke, flames, and forty-degree Celcius heat to stop the fire from advancing. As firefighters battled the stubborn fire that consumed upwards of ten hectares, the women's auxiliary fed firefighters around the clock. At its height, one hundred firefighters, ten pumping units and three helicopters were at work battling the August blaze that at one point got within 150 metres of the Pockwock water-treatment plant.

Lawrencetown Beach Volunteer Fire and Emergency Service

IN 1956 ONE OF THE Lawrencetown Beach area's oldest landmarks, the MacDonald Hotel, was destroyed. The hotel's owner May MacDonald was living alone in the big, old rambling hotel that overlooked Lawrencetown Beach when the fire broke out. Nearly ten years later, in April 1965, the community of East Lawrencetown, on the province's Eastern Shore, had a fire that destroyed its century-old church. Without any organized fire protection there was no hope of saving the building. After these tragedies, the community was galvanized to take action, particularly after an elderly widow's home was threatened by a large grass fire just a few weeks after the church fire. That event pushed local residents to come together to organize the community's first fire department.

The East Lawrencetown Volunteer Fire Department started out with a list of names— nineteen men and two women—of those interested in taking part. With no trucks, no equipment, and no training, the early volunteers had only each other to depend upon. Initially, they paid dues of six dollars a year to be members in the department, which helped to buy equipment. With additional donations from the community and a bank loan, the department was able to buy a trailer, a portable pump, and some hose. Initially, all members had to have trailer hitches on their vehicles so they could be ready to tow the trailer to the fire. The volunteers elected Bob Murphy as their first fire chief.

By the spring of 1967 the firefighters scraped together eight hundred dollars to buy a Dodge truck, with a water tank and pump, from the Waverley Fire Department. That same year, the department built its first station— a two-bay garage—on donated land on Crowell Road. It was eventually replaced in 1991. An additional station was built in 1982 to better serve the demands of the large geographical area. In 1971, the department formed an auxiliary and Linda Macdonald was its first president.

During the 1980s, the Lawrencetown department underwent many changes such as an increasing awareness that as an ocean-side community, water rescue might well be necessary. This became evident one day on the way to a call—the chief had to borrow a boat—which was later returned!—from a local resident's lawn as the department didn't have its own. Without the right equipment, firefighters had to be good improvisers.

Today, the department serves about five thousand residents in an area covering 150 square kilometres of land and water. Though there are some career firefighters, the department still relies heavily on volunteers.

Grand Lake / Oakfield Volunteer Fire Department

THE FIRST MENTION OF forming a volunteer fire department in the Grand Lake/ Oakfield area was at a community meeting on March 11, 1965. By the following June, the Grand Lake/ Oakfield Volunteer Fire Department was formed. Doug Cutler was the first fire chief and Horace Brown was the deputy chief.

The first fire hall was in the basement of the community's old school, which had been turned into a community hall. In order to house the fire truck, a 1958 Dodge with a five-hundred-gallon tank on the back, the volunteer firefighters had to drill and jack-hammer out the cement wall on one side of the building. The truck, which had been built by volunteers, had just enough room to squeeze in. In the early years, the department, like most small rural organizations, scraped by on very little money. In 1967, for example, it had just a thousand dollars to pay for all its expenses and equipment purchases.

In the early days, equipment consisted mostly of second-hand cast-offs from other fire departments. Like other new firefighters, they started out with no breathing apparatus; however, they eventually received some equipment from the Canadian Forces, which enabled them to enter smoking buildings with an air supply as well as some protection from the smoke. In the late 1970s, the department bought a few self-contained breathing apparatuses.

In 1975, the community hall was expanded further to make more room for the fire department. Over the years, the firefighters have continued to work on the building to improve their space, putting in countless hours of free labour. Volunteer members have fluctuated over the years, from a low of six members to a high of twenty-three. Today, the department has about fifteen members and several career firefighters as well.

IN THE MID-1960s, the Johnson family home in Seabright burned to the ground. With no local firefighters to lend a hand, the family had to watch their house burn while firefighters travelled from Black Point more than twenty kilometres away. The fire soon prompted local community members to think about forming their own fire department. In 1967, the community got an old fire truck and stored it at the local sawmill. A few years later they got some breathing apparatuses for the firefighters. Unlike newer apparatuses, when firefighters put them on they could barely hear each other, making communication while on task next to impossible.

Seabright and District Volunteer Fire Department

A fire station was eventually built and an addition; this was expanded to include a second floor and an additional bay for trucks in 1988. Volunteers' skills as mechanics and electricians were necessary to fix any problems with the fire hall or the equipment.

To cope with a lack of fresh water along the rocky coast, the department opted to buy as much high volume five-inch-hose as possible, according to Graham Wambolt, a retired chief with the department. The department now has about 1,980 metres or over a kilometre-and-a-half of hose. This meant that the firefighters could then use a brook, nearby lake, or ocean access from which to pump water. "Usually we would use St. Margaret's Bay if we could access it," Wambolt said, adding that salt water is better for putting out a fire because it is denser than fresh water. However, salt water means the hose and attachments have to be rinsed well afterwards; otherwise, rust is a risk.

Spryfield Volunteer Fire Department. Driver Edward Ingram and caretaker Frank DeYoung of the Spryfield Volunteer Fire Department with the 1937 hose truck purchased for $500 from the Halifax Fire Department, c. 1948

THE HERRING COVE and District Volunteer Fire Department traces its origins back to 1967, two years before the residents of Halifax and its mainland western suburbs underwent a massive change. On January 1, 1969, the communities of Spryfield, Fairview, Armdale, and Rockingham were annexed to the city. Within the local fire departments, eleven firefighters from the Halifax County were transferred to the city department, and another fifty-four were hired, bringing the number of staff at the city's department to 296.

In 1967, the area around Herring Cove was being served by the Spryfield Volunteer Fire Department. Following the annexation, the department was folded. Knowing this was coming, the residents of Herring Cove banded together, and by the spring of 1968 they had their own fire department.

The newly-formed Herring Cove Department started with equipment from the old Spryfield Fire Station and a handful of volunteers from the community. Before long, it grew to become a full-fledged firefighting service with modern equipment and a team of about thirty-two volunteers.

On February 10, 1969, tragedy struck in Spryfield when a fire broke out at a home at 715 Herring Cove Road. When firefighters managed to get inside the building they made a grim discovery: a mother and her seven children were dead. The father, the only surviving family member, was at work at the time.

Tragedy struck again on April 3, 1982, when firefighters responded to an early morning fire at an apartment building at 429 Herring Cove Road. Twenty-five people were rescued but five people died in the fire. Fatally, the building's fire alarm system didn't activate.

LIKE HERRING COVE, the communities of Harrietsfield and Sambro started seriously considering fire protection in their area in 1967. A major fire in the community created concern, as did the knowledge that Spryfield, which the community had relied on for fire protection, was being annexed into the City of Halifax.

Local volunteers got surplus fire equipment, including an old pumper truck, from the Fairview Fire Department. An area rate on the property tax assessment provided initial funding for the department, and the ladies' auxiliary raised more. Firefighters from Spryfield trained the new volunteers.

The pumper was originally housed in founding chief Don Byrnes's backyard; as the weather got colder, it became necessary to store it indoors. Resident Alex Nicholson offered the basement of the house he was building.

L to R: Stephen Brown, Harold Nickerson, Chief Alex Nicholson, Fred Holt, Doug Irons, Captain Clifford Brown, Deputy Chief Bernard Flinn, Dave Dacey, Captain John MacKenzie and Stan Saulnier of the Harrietsfield/Sambro Volunteer Fire Department.

Nicholson's basement soon became Harrietsfield's first fire station. In the early 1970s, construction got underway to build a new three-bay fire station and hall, which was constructed on land donated by Guy Nickerson, next to the Harrietsfield Elementary School. It officially opened in the fall of 1974. To raise money for the station, weekly bingo and Saturday night dances soon became a mainstay of the community's social life. Today, the original fire hall is the site of the Harrietsfield Community Centre.

In the early days, Harrietsfield, like many rural communities at the time, had to rely on fire phones located in volunteers' homes. These members usually lived close to the station and would run over to sound the large siren perched on top of the station. As the community grew, members were notified by phone; eventually, a paging system was bought. Firefighters' pagers were first activated from a base radio station in the hall and later at a base station set up in the chief's home.

As the area grew, a second station was also built in Sambro in the early 1980s. The department's name changed to the Harrietsfield/Sambro Fire and Rescue to reflect the communities it served. Today the department responds to more than two hundred calls a year. Current Chief Bill Powell oversees twenty-three volunteers and three career firefighters.

WITH ONLY GARDEN HOSES and buckets to fight fires, concerns were raised in the community of Dutch Settlement, with a population of about 250, located northeast of Halifax on the Shubenacadie River. In 1969, the first meeting was held to look into the prospect of creating a volunteer fire department. "Everyone was in favour of having a fire hall," said Albert Moore, a founding member and chief of the department from 1976 until 1988.

At the time, the nearest organized fire group was at a factory in Lantz. If necessary, they could be called for aid, but they were never called. After the initial meeting in 1969, several more were held and it was soon decided that a fire hall should be built. In 1971, volunteers undertook the task on a piece of land on the community's main road, Highway 277. The first truck they put inside the hall was a 1952 Ford fire truck, which they bought from CFB Greenwood.

During Moore's years as chief, the department averaged about eight calls a year, for mostly grass and chimney fires. During that time, Moore and his ten volunteer firefighters never had a fire in which lives were lost. "That was my biggest fear—to go out to a house fire and hear someone screaming and they couldn't get out," Moore said.

In the late 1970s, Moore decided the department needed more room in the fire hall. As it was, if they held a meeting in the hall, they first had to move the

Dutch Settlement and Area Volunteer Fire Department

trucks outdoors to make room for chairs. It wasn't an ideal situation because in the winter the water on the truck would freeze. An addition was built on the hall in 1977, making room for an office and kitchen.

THE COMMUNITY OF AFRICVILLE was officially founded in the 1840s in the north end of Halifax, but many of the families who lived there can trace their roots in the area as far back as the 1700s. To some, Africville was a slum, populated by former American slaves who escaped during the War of 1812. But to about four hundred black settlers, it was a place where they could live free from racism and discrimination.

Inside the City of Halifax, the community of Africville posed a serious problem for firefighters, as throughout Africville's history, the city refused to deliver basic municipal services to the community. City staff and aldermen, for the most part, were of the view that rather than extend water and sewerage service to Africville, "the property… be cleared in case some industry might want to go there." But the residents protested, saying that they wanted to stay.

Water or the absence of it was a particularly serious problem; not only did residents lack safe drinking water, but there was insufficient water pressure for firefighters to adequately do their jobs.

A major fire struck the community in December, 1947, and six homes were reported to have been destroyed. The cause of the extensive fire damage was identified as a lack of water supply.

Ten years later, in 1957, another fire in Africville claimed the lives of three children. The absence of a water main and hydrants not only made the

An unsafe water sign posted in Africville, c.1950

community more vulnerable to fire damage, but also prevented residents from getting insurance coverage. In May 1963, a fire destroyed one of the community's best homes. The Farrell family escaped their four-room frame dwelling with only the clothes on their backs after an oil burner flooded. "It was one of the best homes in Africville. He really had it fixed up good. All the houses up here aren't shacks you know," a neighbour was quoted as saying in *The Spirit of Africville*. The *Mail Star* reported that firemen had to draw water from a hydrant more than a kilometre away from the community. They also pumped water from the Bedford Basin in an effort to save the home. "Fire protection adequate? By the time a fire took place out there, by the time the firemen get out there, it was no good; they just watch the place…. No, when a fire started in Africville you just say, give up your home, it's gone, try to get what you can out of it and forget the rest," an elderly resident said in 1969 interview included in the *Africville Relocation Report*.

After the fire, cries from outside the community erupted again, calling on the city to relocate the residents. Africville was described by city officials as a fire trap where tragedy was waiting to happen. Again the residents protested.

In the mid-1960s, the residents of Africville eventually lost the battle to save their community when the city finally decided to relocate them—bulldozers knocked down what was left of the community in 1970.

Chapter 3

The Evolution of Firefighting in the HRM

The pace of change within Halifax and surrounding communities was as rapid in the decades of the 1970s and 1980s as in some of the earlier periods in the city's history. Rural volunteer departments still sprung up but not at the rate they had in the 1950s and 1960s. In Halifax, urban development kept growing with regular office building construction. While the city didn't see the number of skyscrapers rise as did larger Canadian cities, the 1980s saw the construction of two major developments on the Halifax waterfront: Purdy's Wharf Tower I was built in 1984, and Purdy's Wharf Tower II would follow five years later.

New office towers brought new construction technologies—buildings were now sealed and mechanically ventilated. Expanses of mirrored glass, sheet metal, and pre-cast concrete, as well as big, open office spaces with lightweight partitions became the norm. Chemicals and plastics were more widely used in the manufacture of building materials and furnishings.

New construction raised new challenges in fire safety as well. Because it could take at least thirty minutes to empty a twenty-storey building, fire alarms that ordered an evacuation were no longer practical, and some fires were now beyond the reach of aerial ladders. Halifax area firefighters had to be prepared for these challenges, and others. In the 1970s and 1980s there was a greater incidence of both fires and spills of hazardous materials.

Years before Halifax, Dartmouth, Bedford and the rest of Halifax County were amalgamated, a report was written which assessed the overall relationship between local governments and their fire services. According to Kate Birsel's 1984 report, one of the problems identified in the fire services in the County of Halifax was the belief among many of the fire departments that they were autonomous units rather than part of a larger fire service. The report also found that the departments varied widely in size, in the amount of firefighting equipment at their disposal, and in the quality of services they offered.

Birsel found that the annual budgets for the thirty-six individual county departments ranged from one thousand to one million dollars with the total operating budget for these departments at about $2.25 million. Only seven departments had paid employees. In all, there were fifty-seven salaried firefighters, from Cole Harbour/Westphal to Black Point to Sackville. Hundreds more volunteered. Levels of training varied. In 1982, 287 firefighters registered in one or more courses at the Nova Scotia Firefighters School in Waverley. The following year, that figure grew to 335. While the numbers rose, only half of the departments sent firefighters for training.

Vehicles and equipment used by each department were as varied as the communities themselves. "One department may have one truck that is over thirty years old, whereas another department may have just purchased a new vehicle," the report stated, adding that many of the rural departments also had outdated equipment and clothing that could prove to be hazardous.

Response times, which ranged from two to twelve minutes depending on the department, varied because of the different coverage areas of each station. To help, the majority of rural departments were engaged in mutual aid with neighbouring departments. This involved assistance in the form of "standby" or "backup" support in personnel and equipment.

Fire prevention in the rural areas also wasn't consistent. While several departments didn't have fire prevention divisions, they still carried out inspections of houses or other buildings upon request. On the other hand, some of the departments had active fire-prevention divisions. Herring Cove, for example, received national awards in 1980 and 1982 from the Insurance Bureau of Canada and the Fire Prevention Association of Canada for its exemplary fire prevention program.

Meagher's Grant Volunteer Fire Department

LONG BEFORE a fire hall was built and a department formally organized, the community of Meagher's Grant kept a tank at the centrally-located convenience store. In the case of fire, community members would throw the tank on the back of a truck and race to the scene.

On April 23, 1969, a women's group in Meagher's Grant, not far from Musquodoboit Harbour, called a meeting: they wanted to build a new hall that would serve as both a recreation and fire hall, and they needed help. They got it—construction on the hall began later that year, and the first dance in the hall was held on New Year's Eve.

At the same time, the first rumblings for an organized fire department were heard and got underway with an active search for a fire truck. Willis Dickie and Bud Flemming, along with a handful of others in the community, had the job of finding a truck for no more than fifteen hundred dollars. They found one in September, 1970, and next month, the first meeting to discuss the formal creation of a fire brigade was held. The department's first fire chief was Glen Cole, and his son, Cory Cole, is an active volunteer in the department today. The first deputy chief was Ron Dickie.

Though there are only about six active volunteers today, it is expected there will be upwards of ten by late 2005. "Six is barely enough to keep a fire department going," says the department's current chief, Kim Jones. A year ago, the average age of the firefighters was fifty-five, but several have since retired due to age and fitness requirements. Up until the late 1990s, the department still operated with fire phones, as there weren't enough pagers for all the firefighters. They respond to between thirty-five and forty calls a year, half of them medical emergencies. In addition to barn fires in recent years, the department faced two house fires in December, 2004.

Bay Road Volunteer Fire Department

THE BAY ROAD Volunteer Fire Department is located on St. Margaret's Bay Road in Lewis Lake, close to twenty kilometres west of Halifax. It was founded in 1976.

Damage from Hurricane Beth, 1971.

IN 1972, one year after Hurricane Beth—which hit Nova Scotia with 296 millimetres of rain, causing an estimated $3.5 to $5.1 million in flood damage—the Lake Echo and District Department got its start on the Eastern Shore. Members of the community stepped forward to build a one-storey building that would be home to a 1965 Ford pumper.

A few years later, another single-bay station was built in North Preston, a predominantly black community, with Arnold Johnson as the first elected captain. Joan Kennedy is the department's current chief.

Three Harbours Volunteer Fire Department

JUDY SMILEY, a former Halifax County councillor and the wife of Bernie Smiley, the Three Harbours Department's first fire chief, remembers moving to the community of Port Dufferin, east of Sheet Harbour, in 1960. The only fire service in the area was a Department of Lands and Forests truck that travelled from Sheet Harbour. The truck, which had hose that had to be stuck into a nearby brook or stream, couldn't always get to a fire scene in time.

In about 1976, a year before the first meeting was held about forming a fire department for the rural area, Smiley remembers a house in Port Dufferin burning to the ground. It was summer time and tourists stopped to help the residents carry their belongings from the house. "That was a terrible day," she said.

By 1978, the small communities of Port Dufferin, Beaver Harbour, West Quoddy, and East Quoddy had their first fire truck, a 1969 Chevrolet that came from Terence Bay. With no fire hall in which to park the truck, it was backed into the Smileys' yard. In late 1978, Joe Watt, a local resident, donated land and volunteers built a fire hall with money raised through an area rate and fundraisers. The first lobster supper that was held in 1979 to raise money for the firefighters has become an annual tradition. The following year, a brand-new 1979 Ford truck was purchased. Shortly after its purchase, the truck was put to use when a house just outside Port Dufferin went up in flames, the result of a suspected electrical fire.

Bernie Smiley was asked by community members to become the department's first fire chief and he gladly accepted the job, one he held for twenty-three years until he retired in 2000. He remains active in the department's fundraising efforts. "We were one of the few communities around that didn't have a department," said Smiley of the department's start, "we had to make a move."

Unfortunately, the fire chief couldn't keep his family from their own fire. In the 1980s, the Smileys wood/oil furnace caught fire. Before long, firefighters filled their home from basement to attic—being the home of the fire chief, the blaze wasn't taken lightly. Of course, when safety was assured, then "everybody teased Bernie," said Judy Smiley.

For years, the department provided not just fire protection but a free chimney-cleaning service, and, for a small fee, well maintenance. Like most rural departments, Three Harbours has served a dual purpose by both protecting the community, and of almost equal importance, fostering social ties. "It has maintained a nice community spirit," says Judy Smiley. "Everybody in the community supports the whole fire department...the support has never wavered."

Tangier and Area Volunteer Fire Department

ON MARCH 26, 1980 community members held a meeting at the local school to discuss fire and garbage service in the Tangier area. At the time, the residents were said to have been more concerned about getting garbage collection than they were about having a fire department. A fire in Tangier in the early 1980s changed that. The fire left five children dead.

At first the community didn't have a fire truck. Instead, trailers and pumps that were stored in sheds throughout the community were relied upon. When fire broke out, volunteer firefighters would get the trailers and hook them to their cars and speed to the scene. "It was basically a fire brigade at the time," said Chief Darrin Hutt.

On March 25, 1988, council members of the Halifax County Municipality presented the department with a plaque commemorating the official opening of a new fire hall.

The nearby Mooseland Volunteer Fire Department, which got its start in the late 1970s, merged with and became a sub-station of the Tangier and Area department just a couple of years ago.

Mushaboom Volunteer Fire Department

THE COMMUNITY of Mushaboom, located about ten kilometres southwest of Sheet Harbour, was the last volunteer fire department established in Halifax County. Founded in 1989, the department was started in response to the loss of several children in a house fire. Foster Beaver was the department's first fire chief.

Halifax firefighters' strike, 1982.

Both Halifax and Dartmouth firefighters experienced labour disputes in the late 1970s. Dartmouth firefighters were first to walk off the job in a legal strike on January 8, 1979. The 120 firefighters, members of Local 1398 of the International Association of Firefighters, were seeking wage parity with city police officers and protesting a third reduction in wage increases they had negotiated two years previously. It was their first strike. Ron Joyce, president of the union local, told reporters that his association wanted a first-class firefighter with four years experience to earn $19,400 by 1981. At that time a first-class constable on the Dartmouth Police Force was earning $20,280.

By the night of January 10, firefighters had reached a contract settlement, which provided a four-year contract with an increase in the annual basic pay of a first-class firefighter with four years experience from $13,847 to $21,500 per year.

No fire injuries or loss of life were reported as a result of the three-day strike; the only major fire reported in the city was the destruction of a freight shed. A back-up volunteer firefighting force had been on-call in the case of emergency.

Soon after, Halifax firefighters rejected an August 1979 contract offer made by the city. The offer was for a twenty-eight-month contract that would increase a first-class firefighter's annual salary to $19,100 at the beginning of 1981. Firefighters wanted a twenty-two-month agreement that would give them $19,600 by March 1, 1980. At the time, a first-class firefighter earned $15,768 annually. There were 178 first-class firefighters in the bargaining unit of Local 268 of the International Association of Firefighters.

Halifax's 271 firefighters walked off the job on August 17 in the city's first legal fire strike. As union members walked the picket line, a total of thirteen management staff manned two of the city's six fire stations. At the time, each of the two stations had only one truck and one car. Normally, a total of fourteen pumpers and ladders would be available for duty. Volunteer firefighters were not on-call.

Just as Halifax supported Dartmouth during its strike, Dartmouth Fire Chief Bob Patterson said his department would not answer calls from Halifax. Fortunately, the strike didn't last long and was over within four days.

Three years later, on September 17, 1982, Halifax firefighters were on strike again. The city's 242 unionized firefighters walked off the job after Mayor Ron Wallace offered only $24,000 of the $25,760 the firefighters' union had requested. Overtime was also an issue.

Brunswick Street
United Church fire, Halifax,
June 30, 1979

Again, non-union managers manned two of the department's six stations. This time the city organized a group of about one hundred volunteers to patrol neighbourhoods to watch for fires and help to extinguish any blazes that started. Several fires were reported during the strike and while there was property damage, there were no injuries or deaths.

The firefighters stayed off the job until October 20, when they accepted a tentative agreement, with included salary and benefit clauses, as recommended by the provincially appointed mediator Justice Lorne Clarke. In July 2004, the city made a deal with its firefighting union to protect the public against any walkouts or wage disputes for twelve years. The new contract, which is effective until May 31, 2016, will see firefighters' wages increase gradually. The first raise, increasing a first-class firefighter's salary from $52,000 to $55,000, took place in 2004. First-class police officers earned about $62,000 at the time. The disparity between a firefighter's salary and that of a police officer will be reduced to about five percent by the end of the contract. The new contract will increase the $31.5 million spent in operating costs in the region's core area by about 3.4 percent every year over the next twelve years.

Sir James Dunn Law Library fire, 1985

CLEANING STAFF noticed smoke and sparks on the fifth floor of the Sir James Dunn Law Library at Dalhousie University about 7:00 A.M. on August 16, 1985. The library occupied twenty-four thousand square feet of space on the fourth and fifth floors of the Weldon Law Building.

Fire engines arrived on the scene minutes later, but the top floor of the building was already engulfed in flames. Fire raced through the library, destroying approximately sixty thousand of the library's 160,000 books and periodicals. By 9:00 A.M., firefighters had extinguished the flames, but the roof had caved in after steel girders buckled from the heat. The cost of the clean-up was estimated at $250,000. Losses were in the millions.

The fire, which was caused by lightning, sparked debate over the need for sprinkler systems. The law library lacked sprinklers, having been built in the late 1960s, before provincial regulations were brought in which stipulated that buildings of the library's size must have sprinklers.

Chief Director
Michael Eddy in 2001

IN 1977, at age twenty, Michael Eddy joined the Dartmouth Fire Department. While his father wasn't a firefighter, his grandfather had fought fires underground in the Springhill mines. Eddy had initially set out to become a policeman, but that department wasn't hiring and suggested he try the fire department instead.

The weekend before he started with the Dartmouth Fire Department, he had to make a trip to the barber for a hair cut; at the time, firemen were being fined if their hair was too long, he recalled. He had to visit the barber a second time because his hair was still considered too long after his first visit.

Since Eddy's home was in Eastern Passage, he also volunteered with the Eastern Passage/Cow Bay Department. By 1991, he had been promoted to captain with the Dartmouth Department, and two years later, became fire chief for the Sackville Department. The following year amalgamation was announced, and on January 1, 1996, Eddy began his new job as deputy commissioner of fire. With Commissioner of Fire Gary Greene, a twenty-year veteran of the Dartmouth force, and Bill Mosher, Deputy Chief, Administration of Halifax Fire, he set out to create a new amalgamated fire service.

THE DARTMOUTH FIRE DEPARTMENT experienced a changing of the guard in 1987 when Chief Robert Patterson retired. He would leave the department with 131 career members and eighty volunteers. Patterson had not only been chief of the department since 1978, but a member for thirty-eight years. His father George Edward Patterson had been the new City of Dartmouth's first fire chief.

"I'm not that old but I can remember the last couple of horses Dartmouth had to pull its fire engines," he told a reporter in February, 1987. "I can still remember chasing after them when I was just a little fellow."

Fire consumed St. Peter's Church on Maple Street in Dartmouth, Dec. 28, 1966, one of the largest fires in Robert Patterson's many years with the department

Patterson recalled that when he was hired in 1950, he was only the twelfth man to be hired permanently by the Dartmouth department. Firefighting in those days wasn't easy, he said. The department was making the slow changeover from the old soft rubber hose, which was very heavy and cumbersome, to more modern hoses made of a lighter material.

Recalling old-time stories, Chief Patterson said that he received his worst injuries not in a spectacular blaze, but during an ordinary garage fire. After neglecting to pull his boots up all the way to protect his legs, some flying ash caught his clothes on fire, and he was badly burned from the tops of his knees to the bottom of his coat. "There were also a few instances with the Natal Day fireworks which gave us a close call. One time one of the fireworks didn't go off correctly and it blew the laces right out of my boots," he recalled.

ALONG WITH AN INCREASED awareness and focus on fire prevention, the 1980s marked the beginning of a new era of concern over hazardous materials (known as HazMat). As firefighters are often the first to arrive at the scene of an accident involving hazardous materials, they needed training to protect both themselves and the public. In the 1980s, the Halifax Fire Department took advantage of provincial and federal funding to buy a vehicle with specialized HazMat equipment, and HazMat teams were set up throughout the region. Halifax's team got its start in 1986.

Hazardous Materials Teams, 2004

The following year, the department's training was put to the test. Newspaper headlines on April 2, 1987, blared: "Gas Scare Causes Evacuation of 250 Homes." A leak in the gas storage facilities of a Halifax service station in the Chebucto Road and Elm Street area caused the problem. Firefighters, with the help of the company involved, worked for days to solve the problem. A month later, a potentially explosive situation occurred at the Technical University of Nova Scotia when four propane tanks were knocked over and started to leak. Police sealed off the area while firefighters prevented a disaster.

Today, the fire department and its decontamination vehicles, which carry a massive tent capable of decontaminating large numbers of people, are ready for almost anything. There are ten members of the HazMat team at the Highfield Park Station and another ten members at Station 5 on Bayers Road. Trained members are also scattered throughout several other stations. The fire station on Highfield Park Drive in northend Darmouth is considered the department's main HazMat station. The station includes an entire decontamination wing with a series of rooms for removing contaminated clothing, as well as laundry and shower facilities.

AN INTEGRAL PART of the regional fire service, dispatch has evolved from ringing bells on top of fire halls, to fire phones in the homes of firefighters, to box alarms, to computer-aided dispatching.

In 1987, the fire department first hired dispatchers whose sole job was to transmit alarms for the fire service. Prior, firefighters were required to take turns in the dispatch centre. In 1997, a joint police and fire dispatch centre was set up at the Bedford Fire Station. As of July 7, 1997, residents in Halifax, Dartmouth, and Bedford could dial 911 to get police, fire, or ambulance emergency response. Emergency calls from all other communities in the municipality are first rung through to the RCMP's centre on Oxford Street in Halifax and are then transferred to the dispatch centre. However, this process is changing—in 2003, the rural departments within the municipality agreed that the dispatch centre should handle their calls.

HRM Fire Dispatch Centre

In the fall of 2005, a new 911 dispatch centre is scheduled to open at the Eric Spicer Building in Dartmouth. The new centre will be equipped with a new computer aided dispatch (CAD) system that uses computerized radio equipment and content knowledge. The CAD system will automatically tell a dispatcher what equipment should be sent to an emergency scene.

An essential part of dispatch is communication between fire trucks and the 911 centre. In 2000, the fire service brought in a trunked mobile radio system for rural areas, which provided better communication and radio coverage to the municipality's more remote areas. Management decided to give the rural stations the new system first as they appeared to be in the greatest need. In some areas, such as the Musquodoboit Valley, firefighters were unable to communicate with dispatch once they left the station. Cellphones didn't work in the area, making communication next to impossible. The fire service's urban core stations got the new system in 2003 and 2004.

Sixty people currently work in the Bedford dispatch centre. Twenty-four-hours a day, three dispatchers are dedicated to fire calls. Close to seventy thousand 911 calls come through the centre a year. Of those, about eighteen thousand are for firefighters.

When a call comes in about a structure fire, motor vehicle accident, hazardous spill, electrocution, or someone having serious breathing troubles, the dispatcher automatically sends a fire truck to the scene. For other medical calls, the Emergency Medical Services (EMS) dispatch centre in Dartmouth decides whether firefighters are needed. They are often dispatched to provide back-up or extra support such as helping to lift a patient.

In the rural areas dispatch works a little differently. Each department has an agreement with EMS as to which types of medical calls they will respond to. In some smaller areas, the department only responds to life-threatening calls, whereas in more suburban areas, the departments tend to respond to a broader range of calls.

At one time, firefighters were spending seventy percent of their time responding to medical calls. In the last few years, Mike Mahar, retired manager of the 911 Communications Centre, and senior fire officials met with medical officials in the government and other stakeholders to determine how the fire department could cut back on the number of medical calls, as it became clear that sending an ambulance with two paramedics to a scene rather than a large fire engine with four firefighters was not only more economical, but a better use of resources.

Dispatching for fire involves much more than it does for police, according to Mahar. For example, if there is a structure fire, the dispatch centre not only sends out the first trucks but will manage all the subsequent calls that go out to the chief, Nova Scotia Power (if power has to be cut off), the water commission, fire prevention, and for any specialized equipment. For example, a structure fire in the woods near Terence Bay a few years ago required eighteen trucks to be dispatched to the area.

Chapter 4

Many Communities, One Department:
The Impact of Amalgamation

The 1990s brought tremendous change for the department, the biggest of which was municipal amalgamation in 1996. Residents, for the most part, greeted amalgamation initially with fear and loathing—they wanted to retain their own communities and not live in a municipality made up of more than two hundred urban, suburban, and rural communities and nearly 350,000 residents. Amalgamation was also not embraced by the region's fire departments. At the time, the fire service went from sixteen hundred firefighters operating in over thirty independent fire departments in a region spanning 5,577 square kilometres into one regional service. Gary Greene, Commissioner for Fire Services for the Halifax Regional Municipality at that time, understood the challenges that lay ahead. In March 1996, he told the *Chronicle Herald* that it would be 2010 before everything really settled down.

Years before amalgamation became a reality, the fire service had more pressing issues to deal with. One of them occurred on August 22, 1991. The National Film Board (NFB) Building, which originally housed the St. Mary's Young Men's Total Abstinence and Benevolent Society, was destroyed by fire. Built in 1891, the building featured an unusual convex mansard roof and was part of the historic Victorian streetscape on Barrington Street that included the City Club and the Bean Sprout building. Along with the four-storey building went the NFB's production facilities, a film and video library, administrative offices, the Atlantic Film Festival's office, and a theatre. At least ten neighbouring businesses were also affected by the fire.

ON JUNE 2, 1994, at about 5:30 P.M., one of the most dramatic fires in modern Halifax's history took place. Three young boys were seen entering a side door that led to the basement of historic St. George's Anglican church on Brunswick Street. They were later discovered to have been playing with matches. Minutes later, a parishioner who lived nearby was reported to have seen smoke coming from the basement window and called the fire department. Soon, fire was burning in the women's choir room on the south side of the basement under the front porch. The open stairways and concealed spaces behind the walls allowed the fire to spread to the top of the porch and into the ceiling space above the umbrella dome. There, it ignited the framing timbers and found pathways to the ceiling space over the chancel. Once over the organ loft, the fire spread into the church from the opening behind the organ pipes, as Elizabeth Pacey describes in her book *Miracle on Brunswick Street*.

Firefighters flooded the scene to battle the flames. Smoke was visible throughout the city as the historic structure, built in 1800 on orders given by the Duke of Kent, burned. In total, thirty-two firefighters spent more than twelve hours pumping four thousand gallons of water per minute through their hoses. Many doubted the church would survive. Parishioners wept and prayed over the loss of their house of worship, known as the "Round Church." Some of those prayers were for firefighters who battled the blaze.

Fire engulfs St. George's Round Church, 1994

The flames eventually engulfed the cupola and sent up a cloud of black smoke. When the cupola collapsed, cries of horror could be heard from the crowd of helpless onlookers standing nearby.

St. George's Church was the house of worship for two hundred families and 150 churchgoers. Greg Videtec was one of them. He attended evening prayer just before the fire began. "We're absolutely sick at heart," he told the *Daily News*.

"I've had better days," Robert Martin, a church warden, told the newspaper. "Even if we build a new church, we can't replace this church because it's unique."

Following the fire, parishioners organized a massive fund-raising campaign to restore the building to its original condition, at a cost of about five million dollars. At 5:30 P.M. on June 2, 1999, exactly five years after the terrible fire, a Festival Evensong service was held as a thanksgiving tribute to the twenty-five hundred donors, the dedicated workers, and the many volunteers who made the restoration possible.

Less than six months after the Round Church tragedy, the fire department was forced to turn its focus to amalgamation plans. In October 1994, the province decided to proceed with amalgamation of the City of Halifax, City of Dartmouth, Town of Bedford and Halifax County Municipality. At a meeting a month later with then Mayor Randy Ball, fire chiefs from throughout the region raised their concerns about amalgamation and how it would affect their volunteer fire departments. Fire service in rural communities is "an essential community organization providing much more than firefighting services. In many cases the fire hall is a centre of community activity and of great importance to the general well being of the community," they told the mayor, according to minutes taken from the meeting.

The fire chiefs also pointed out that their rural departments had very little in common with the fire departments in Halifax and Dartmouth. They noted that, as a whole, the county's fire service was not only larger than the two city departments, but more cost- efficient because of being largely volunteer. Having been run independently for so long, the chiefs worried about how they would operate after amalgamation. They wondered if their departments would continue as volunteer services or if they would become part of a new fully paid service. Many of the concerns raised by the fire chiefs at the meeting would play out not only in 1996, when amalgamation officially took place, but for years afterwards.

Impact of Amalgamation

Gary Greene, the first fire commissioner of the newly created Halifax Regional Municipality, 1996

HRM Emergency Measures Organization. Barry Manuel (at right) sits in on a meeting at the organization's office in Dartmouth, c. 1996

ON APRIL 1, 1996, the Halifax Regional Municipality was officially established. The Town of Bedford, City of Halifax, City of Dartmouth, and Halifax County amalgamated to form a municipality covering an area larger than Prince Edward Island. Months before the new municipality was officially formed, the hard work of creating what would become Halifax Regional Fire and Emergency began. What lay ahead was the challenging task of bringing together the thirty-eight independent fire departments within the municipality.

Bill Mosher, the current deputy chief of rural operations, remembers reporting to the Maritime Centre for work in early January 1996. Mosher and fellow deputy commissioner Michael Eddy arrived at the office along with Gary Greene, the new commissioner for fire services for Halifax Regional Municipality. The two deputies found desks waiting for them but no phones or computers.

"Our job was to build a fire department," Mosher said. They started out by creating a structure for the new department, drawing up budgets, and hiring staff. Work also began early on to bring the reluctant rural fire areas onboard. "The rural departments were so scared that their equipment was going to be taken away from them and taken to the city—equipment they had worked so hard for," Mosher said. "There was a lot of resistance at first." Particularly strong opposition to amalgamation came from the departments in Herring Cove, Musquodoboit Harbour, Bay Road, Seabright, Oyster Pond, and Beaverbank, said Mosher.

"It was quite a challenging period," Eddy said. "There were some long, long days."

WITH AMALGAMATION also came the creation of the Halifax Regional Municipality Emergency Measures Organization (EMO). Working closely

Black Point Volunteer Department's 49 engine, c. 1960

with the provincial EMO, the organization would become HRM residents' direct contact in the event of a large-scale emergency or disaster in their area. A year after amalgamation, former police officer Barry Manuel took over the regional EMO. Manuel, whose job falls under the responsibility of the fire services, describes his job as that of a facilitator: "I make snowballs. Other people throw them," he said.

In a large-scale emergency where various agencies such as fire, ambulance, police, and public works are involved, Manuel works to ensure that a lead agency is established at the scene and that everyone is working together. The job can be a massive undertaking. He describes one particular Victoria Day weekend forest fire in Eastern Passage that required the evacuation of up to seven hundred people. The evacuation process, along with fighting the fire, required the involvement of seventeen public and private agencies. Manuel's biggest test came just a year after he was on the job, with the Swissair crash of 1998.

IN THE DECADE following amalgamation one of the biggest challenges facing the new regional fire service was uniting an urban paid service with one which was rural and largely volunteer. Fire officials met a lot of opposition along the way. Traditionally volunteer fire departments have symbolized autonomy from government and strong identification with their community. These departments are often regarded as the pillars of the community. The stations are usually situated in a central location in their communities with each station either affiliated with the community hall or acting as the community hall.

With amalgamation, the formerly independent fire departments in the communities around Halifax and Dartmouth merged to become part of HRM's bureaucracy. Hierarchies, procurement processes, and regulatory controls all changed, and in the process more demands were put on volunteers' time.

"This radical shift in organizational culture has been met with high resistance from the volunteers. In particular, the volunteer chiefs generally perceive the benefits of amalgamation with skepticism, and associate the change with a significant loss of autonomy and status," says a 2001 report called *HRM Fire and Emergency Service: Rural District Fire Service Management Review and Recommendations*. "The current situation is having an adverse effect on the quality of the organization's relationships and communication processes."

The communities outside the urban core were divided into municipal electoral units headed by county councillors before amalgamation. Each municipal unit set its own tax, or area rate, based on the needs of the community. The volunteer chiefs and municipal political representatives enjoyed a good deal of autonomy. Typically, the fire chief negotiated the fire and emergency portion of the area rate with a councillor. Once the rate was approved, the fire chief would assume full responsibility for the planning and disbursement of the tax revenues. Most departments did a lot of fund-raising to supplement the tax revenues. Due to this funding model, departments could take real ownership in their equipment and hall, often viewing it as their own property.

After amalgamation, the size of the expected budget reductions, about four million dollars, forced management to focus its attention on the paid staff in the urban and suburban areas rather than on the volunteers. Negotiations to reduce the size of the career firefighters were complex and full of conflict.

In late 2000, the fire service created rural district chief positions, a move which upset several rural chiefs. The position was added to provide administrative support to the volunteer chiefs struggling with an added workload. District chiefs also became responsible for safety, training, fire prevention, policy reviews, communications, and procurement.

Another change that came with amalgamation was the way in which chiefs in the volunteer department got their posts. Prior to March 1998, composite and volunteer chiefs were elected to their positions by members of the department. However, around that time HRM started selecting candidates and paying the composite chiefs a modest stipend for their services. The composite chiefs, who are responsible for both paid and volunteer staff, now tend to work in the busier, more suburban areas, such as Hammonds Plains, Waverley, and Prospect Road. Composite chiefs receive an annual stipend of anywhere between seven and fifteen thousand dollars.

A new system is also now in place where every volunteer firefighter in the rural areas receives an honorarium based on a point system. Annually, they make at least $750. It is a small amount of money, considering how much time they invest and how often personal vehicles are used to travel to a fire scene.

Tanker Shuttle in Tangier, 2003

DEPUTY FIRE CHIEF Bill Mosher doesn't hide his respect for rural volunteers, stating that "they're the most valuable resource this community has." In an increasingly busy world, many volunteers in the rural departments continue to put as much time into the fire department as they do their regular jobs, says Mosher.

Today, volunteer firefighters easily put in more than two hundred hours a year, half of which is spent in training and meetings. "Volunteers will be just as important in the future because in small communities, it's not economically feasible to do it any other way," says Bernie Turpin, a district fire chief with the fire service. The challenge in the future will be in finding enough volunteers who are willing and able to make that kind of time commitment to their community. Age is also a consideration, as volunteers aren't getting any younger. In 2001, a survey found that just over forty-one percent of respondents were over the age of fifty, and the average length of service in the departments was just over fourteen years.

Mosher remembers waking up Christmas morning in 2004 and learning that members of the Black Point department were not at home with their families opening presents but out in their fire boat helping someone in need. The following day there was a house fire in the small community of Middle Musquodoboit. It was snowing and cold and the volunteers weren't getting paid but still they were out trying to save the house. In Beaverbank, firefighters were also out battling a blaze that Boxing Day. "The system we have today is the best one because the chiefs take ownership of their communities," says Mosher.

Despite the praise from Halifax Regional Municipality Fire and Emergency's top brass, some of the rural departments weren't always so sure they wanted to be part of the amalgamated regional service. In the fall of 2000, the volunteer departments in Musquodoboit Harbour, Oyster Pond/Jeddore and Beaverbank/Kinsac threatened to separate from the regional fire service. They said they were fed up with having to adopt what they called urban-oriented policies. One particularly upsetting issue for the departments was the chief's plan to assign four new district chiefs to the rural areas. The volunteer firefighters saw that move as meaning a boss was being sent from the city to watch over them. In the end, the departments didn't follow through on their threats to separate.

On April 21, 2004, the department displayed ten new fire trucks at the Woodside Ferry Terminal. Eight tanker/pumpers are headed to the rural fire stations, and two engines will stay in the city. The new apparatus are part of an ongoing effort to upgrade equipment in rural areas.

Since amalgamation, the regional fire service has been standardizing equipment in all its stations. The rural departments all eventually had their budgets taken over by the regional service, and in the early 2000s, a new tax structure came into place, bringing equal funding across the board. Instead of one department having a 1970s truck while another had the most modern piece of equipment, the new funding model meant the fire service could work to ensure the same level of equipment across the region.

In June, 2002, the Halifax regional fire service announced its plans to buy thirteen new trucks which were sorely needed at stations that had been working with aging equipment for too long. Most fire trucks need to be replaced about every fifteen years because technology becomes outdated and

Upgrading the fire service's equipment, 2004

maintenance costs increase sharply. Topping the list of the purchases was a thirty-three-metre aerial ladder truck, worth about $800,000 fully equipped. It was considered to be the tallest aerial truck in the Maritimes at the time.

Four pumper trucks, worth about $450,000, were headed to suburban fire stations in Hammonds Plains, Eastern Passage, Cole Harbour/Westphal and Middle Sackville. Each truck could hold eight to nine thousand litres of water. Four tanker trucks, each worth between $225,000 to $245,000 and capable of carrying about seven thousand litres of water, were sent to other departments. Two were headed for the Fall River/Waverley area, one to Herring Cove, and one to Lawrencetown. Two rescue pumpers, each worth about $275,000, were headed to Oyster Pond and Beaver Bank. A large pumper truck, worth about $300,000 and capable of carrying about eleven thousand litres of water, was headed to Middle Musquodoboit to serve the Musquodoboit Valley, and a rescue truck worth about $225,000 was sent to Musquodoboit Harbour. Since the rural areas are broken up into zones, the idea is that the departments in a zone work together so that they can share access to the larger, more expensive pieces of equipment.

In addition to the new trucks, another $488,000 was spent in the years after amalgamation to upgrade breathing apparatus, giving each rural department the most modern equipment, says Turpin. Another $400,000 was also spent on upgrading water supply systems in the rural communities, through items such as dry or non-pressurized hydrants.

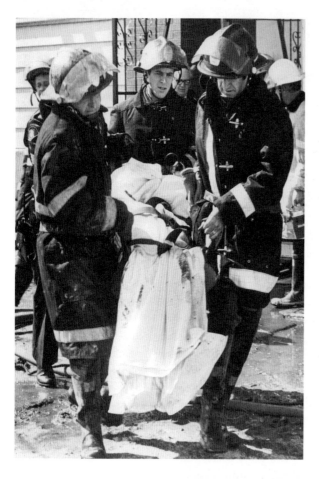

In the line of duty

MOVING INTO THE FUTURE, Mosher expects funding will be the biggest issue facing rural stations. Since several rural communities built their own fire departments with limited funds between the fifties and seventies, the halls are small, aging, and without proper facilities. Upgrading the halls is a pressing issue. About $300,000 was spent recently on repairs in Cooks Brook after mold was discovered in the station. In Moser River, a new fire hall is being built and is expected to open in the late fall of 2005. In the Fall River/Waverley area, a site for a new joint fire station and recreation centre has been chosen, and construction is expected to begin in the spring of 2006. Major deficiencies in fire halls in the Muquodoboit Valley and other areas have been identified and upgrades or new stations will be built as funding becomes available, says Mosher.

Just as the process of standardizing equipment, apparatus, and fire halls across the region was taking place, the regional fire service also turned its attention to standardizing response times. It developed standard responses for rural, suburban, and urban areas. For districts with a population density over one hundred people per square kilometre, a dispatch time of sixty seconds was established as a standard by HRM with a response time of five minutes or less for the first truck to arrive at the scene. For districts with a population density under one hundred people per square kilometre, the standard dispatch time is sixty seconds with a response time of ten minutes or less.

As the rural and suburban communities around Halifax and Dartmouth grow and the pressures on municipal services increase, the fire department adapts by placing paid staff in a once all-volunteer department to ensure the community's safety. For example, in Hammonds Plains, this eventually happened after the community experienced about three or four calls during the day and none of its volunteers could respond, which necessitated that Bedford, the closest nearby department with paid staff working at the station, responded to the calls instead. In a further effort to standardize service, Chief Director Michael Eddy is pushing to have a new standard set so that when the population of an area reaches a prescribed level, paid staff will automatically be added to a fire station to augment volunteers, especially during day-time responses. The industry standard indicates that four firefighters must be at a scene before anyone can enter a building with equipment.

BARB SAWATSKY has come a long way since becoming a volunteer with the Prospect Road and District Fire department in 1991. Having just moved to the area, she wanted to meet people and feel a part of the community. Not interested in quilting bees or baking for social events, she took the advice of a friend in the fire department and went down to the fire hall. During her interview at the Hatchet Lake station, the then thirty-two-year-old dental technician remembers being asked: "Are you afraid of blood? Do you faint at the sight of blood?" When she answered "no," she was handed a pair of oversized bunker gear and boots and was told, "You're a firefighter now." No physical test was required.

Today, Sawatsky is chief of the department and in charge of forty-five people, six of whom are paid staff and the rest are volunteers. She is also part of a growing number of women who are slowly finding acceptance in a male-dominated profession and changing the face of firefighting in the process.

In the early 1990s, there were two women in the Prospect department, but they did mostly office duty and less, if any, time on the front lines. About six months after joining, Sawatsky was given a pager and told she could respond to calls. A couple of humiliating experiences taught her that if she was going to do the job she had to learn to do it right. The first such experience was during a call in the middle of winter to a bush fire in Terence Bay. Sawatsky, who has always been physically strong, remembers trudging through knee-deep snow in her oversized gear and struggling with the heavy equipment on her back. She was with another struggling female firefighter. At one point she remembers falling waist-deep into the snow. One of the male firefighters coming up from behind gave her a hand and bluntly suggested that they stay behind. "We'll pick you up on the way back," he told her. "We were just so embarrassed," she recalled.

After another similar humbling experience she sat herself down. "I decided if I'm going to be here I'm not going to give up my axe." It has since become her motto, especially with new female recruits.

Women on the Halifax Regional Fire and Emergency Service team L to R: Patty Dunbrack, Andrea Sparanza, Denise Patey, Heather Richards, Samantha Meehan, Mariane LaPointe

Sawatsky not only had her physical prowess challenged in the early days, but she often faced condescending attitudes from her male colleagues. At the scene of a motor vehicle accident in the mid-1990s, she remembers an officer handing her some gear. "Sweetheart, will you put this back on the truck?" he said. "I was really PO-ed. I went to the chief," she recalled. "I said call me firefighter, call me Sawatsky, call me Barb but I don't appreciate being called 'sweetheart'." The chief called the officer into his office. After that incident, the male firefighters in the station tiptoed around her for awhile until she broke the ice. She told them that if she was going to wear the same gear as them and do the same job all she wanted was to be treated with the same respect.

Joan Kennedy, Lake Echo Fire Department's current chief, told the *Chronicle Herald* in 1999 that when she joined the fire service twelve years earlier, she raised eyebrows and faced low expectations from men who didn't expect her to last. But she didn't quit and says she was driven by her love of looking after others. "You don't worry about the little words and the little looks you get, that's nothing. You've got to let it bounce off you and not take it personally. I'm a go-getter and a workaholic and my biggest thrill is to give to the community and people."

Debbie Gaudet was the region's first paid female firefighter, hired in Dartmouth in June, 1991. She has since left the service. Today, the regional fire service has seven paid female firefighters and seventy-five volunteers. In the fall of 2005, the department plans to launch a designated recruitment drive to attract more women into the fire department. Sandi Vidito, the fire service's deputy chief of support services, said she isn't setting a recruitment target but will instead respond to the demand. As part of the recruitment drive, some of the service's current female firefighters are heading out into the community to speak about their experiences, life in the fire hall, and the benefits of being a firefighter. "Women now in the department say things are gradually changing," says Vidito, who joined the fire department in 1996, becoming one of the first female deputy fire chiefs in the country. "You don't have to be one of the guys; you can be your own person."

Increasing diversity in the fire service
Clockwise: Corey Beals, Kevin Reade, Jeff Paris, Blair Cromwell, Cyril Fraser, and Steve Simmons

THE FEMALE RECRUITMENT drive of 2005 follows in the footsteps of a successful 2003 drive aimed at getting more black firefighters in the service. "We noticed that we weren't getting a good representation of the community," said Vidito, adding this included women and visible minorities.

The fire service developed the recruitment program after working with community groups such as the Black Community Workgroup of Halifax Co-operative Limited, Watershed Association Development Enterprises, African Canadian Employment Clinic, and YMCA Enterprise Centres. Since 1996, the fire service had been trying to attract more African-Canadians to train as firefighters. Over the years, information sessions have been held at the Black Cultural Centre, and the fire department worked with the African-Canadian Employment Clinic, as well as developing new hiring and work placement programs. In 2004, the work paid off when the fire services hired eleven new black firefighters. Even with the new hires, however, fewer than twenty of the fire service's 426 career firefighters are black.

Chris Powell always wanted to work in emergency services—first as a police officer, next as a paramedic, and eventually as a firefighter. In 2004, at the age of thirty-two, he finally got his chance to live out his dream. "It is a big adrenaline rush," Powell told the *Chronicle Herald* in 2004, describing the excitement of answering a fire call. Ray Adekayode, another of the new black recruits, said he hoped to help the black community by inspiring others to become firefighters. "I'm very proud of the job and it's a great opportunity for me to [be a] role model," he told the *Chonicle Herald*. Firefighting is a job that seems to sometimes attract families, with children following in the footsteps of their parents, he said. Adekayode speculated that may explain why black Nova Scotians haven't traditionally been attracted to the job—they didn't see anyone like them in the department.

THE DEPARTMENT has come a long way since Dartmouth resident Walter Brown became a volunteer with the Dartmouth Axe and Ladder Company and the Dartmouth Union Protection Company, in about 1920. He was sixteen. He wasn't a member of the department, but when he heard the fire bell ring, he would pull on his ankle-length rubber jacket and boots and come running to help drag hoses and help out however else he could. According to one account, he reportedly said that he was only given demeaning "go-for" jobs and little respect by the firemen. But on December 24, 1927, the members of the fire company presented him with a watch for all his hard work and support.

Walter Brown

Chapter 5

Noteworthy Fires and Disasters
from the Last Thirty Years

A call came in to Halifax's Search and Rescue Co-ordination Centre on September 2, 1998, from the Moncton-based air traffic control centre, telling them a large airliner had disappeared from radar. Within minutes, HMCS *Preserver*, Canadian Coast Guard vessels, and rescue aircraft had been dispatched to the site, where Geneva-bound Swissair Flight 111 was believed to have gone down.

At 10:31:22 P.M. on September 2, scientific instruments at the Bedford Institute of Oceanography near Halifax recorded a "peak" seismic event coming from the direction that would later be identified as the crash site. The jet crash off Peggy's Cove killed all 229 people on board.

The community of Blandford, located in the next-door municipality of Chester, was the first department dispatched to the crash site. With the threat of large-scale fires, they made a request for foam to nearby departments. Seven of the area's fire departments are reported to have responded to the crash within the first thirty minutes.

"We initiated mutual aid with departments in Chester, Chester Basin, Hubbards, and Black Point. We requested pumpers, rescue vehicles, and all the foam we could get because we didn't know if it was going to be on land or water," Chief Phillip Publicover of the Blandford Volunteer Fire Department told *Fire Fighting in Canada* magazine in 1999.

Ten kilometres from the Blandford station in Aspotogan, opposite Peggy's Cove, there was a strong smell of fuel in the air. Emergency crews searched the area, but found no debris. It soon became obvious that the flight had gone down not over land but in the water. Fire trucks from volunteer brigades all over St. Margaret's Bay assembled at Bayswater Beach to await further instruction; the beach became the initial staging area for operation headquarters. Firefighters soon started calling anyone they knew who had a boat in case they needed to do a water search.

Before long, Blandford's Deputy Chief Jim Hatter was searching the water in a six-metre boat. Other firefighters joined in a larger search. The RCMP took charge of the scene, assigning firefighters a number of tasks such as security and setting up a first-aid area should they find survivors.

Overhead, Canadian Forces helicopters circled dropping flares in an attempt to establish a crash site. By morning, authorities had determined that the plane had gone down about three or four kilometres from Blandford. Despite the crash site being so close to Blandford, the RCMP decided to move their command post to Peggy's Cove, as the popular tourist destination provided three key elements: one road in and out of the tiny fishing village so access could be controlled; amenities at the village's Sou'Wester Restaurant; proximity to the Halifax-area infrastructure needed for moving equipment and supplies. After the command centre was moved to Peggy's Cove, Blandford, the first fire department to respond to the call, had to step aside, as Peggy's Cove is in the jurisdiction of the Seabright Volunteer Fire Department.

Barry Manuel had assembled a team consisting of District Fire Chief Bernie Trupin, who was then volunteer fire service co-ordinator and a volunteer firefighter with the Blandford department; Chief Zwicker,

Fires and Disasters

Firefighters are illuminated by the lights of a Labrador Search and Rescue helicopter, as it scours the shore of Peggy's Cove, looking for survivors and bodies from the downed Swissair Flight 111.

who was responsible for fire suppression; Chris Charron, then HRM Fire Services logistics manager; Chief Douglas Avery, then of the Prospect Road and District Fire Department; and Chief Terry Stewart, then chief of the Lakeside Fire Department. "Getting the infrastructure and roles established is half the battle," Avery told *Fire Fighting in Canada*. "The first four days were very intense and hectic. Once everything was in place the infrastructure kicked in and worked really well, we were in for the long haul."

After an EMO command centre was set up, establishing a communications system was the next priority. The fire service had its own system using radio and cellphones. Ham- and short-wave radio operators were brought in by the EMO for ship-to-shore communications.

The Seabright fire hall became the headquarters for co-ordinating the fire services' manpower. Within seventy-two hours of the crash, an estimated 240 firefighters from forty departments responded to the scene. "Volunteers were working their day jobs and were spending their evenings relieving people on the job," Turpin told *Fire Fighting in Canada*. He was one of many who logged twenty hours a day for the first five or six days after the crash.

Graham Wambolt, a former captain with the Seabright department, was one of the volunteers who took time off from his paying job as a mechanic. Along with the help of other firefighters, his initial task was to set up lights and generators for a secure helicopter landing area in the back parking lot of the Sou'Wester Restaurant. When hordes of media started arriving the morning after the crash, the firefighters were in charge of managing their arrival and movement. "The organization of the whole incident went well," Wambolt said. "There was good co-operation between all the emergency services." As time went on, as many as four hundred volunteer firefighters from across the province came to lend a hand. They helped support agencies such as the Red Cross in whatever ways they could. "They were all fantastic. I can't say enough about them all," Avery told *Fire Fighting in Canada*. "Their generous nature and caring attitude exemplified the word volunteer."

As it became clear that there were no survivors from the crash, the fire department helped to set up a decontamination centre for handling recovered body parts and pieces of the wreckage. Organizing the centre meant quickly gathering everything from rubber gloves to lights for the search boats. During the actual recovery stage, firefighters took over the decontamination process, which meant washing down recovery workers with chlorine and water.

The hardest part came when the crash victims' families started arriving. "Firefighters were used to responding to emergencies, analyzing the situation and dealing with it but when the families of victims started arriving and you begin to talk to them and they become real people, that is the part that brings it home to you," Turpin said. One of the jobs firefighters and other emergency workers found themselves doing was helping devastated family members lay wreaths in the ocean for those they lost.

Teamwork and ingenuity were key elements needed to cope in the immediate days after the Swissair disaster. The scale of the disaster meant that even the best laid plans sometimes couldn't match what unfolded in reality. This happened when Red Cross Director John Byrne passed along a request for some refrigeration trucks to Barry Manuel. Initially Manuel didn't think the request would be a problem. He thought all he'd have to do was go to his trusty resource book and find the section labeled refrigerator units. Over the years, Manuel had carefully compiled a list of every type of resource he thought he might need in an emergency, but there was nothing for refrigerated vehicles. "What saved me," he said, "was a firefighter standing outside my mobile command bus who overheard me say, 'Where am I gonna find five reefer trucks?' 'Simple,' said the firefighter. 'My brother works at Agora Foods. I'll call him. He'll have their supervisor call you.'" Within an hour, Manuel marveled at how the company had dispatched five trucks to Shearwater, where a temporary morgue was being set up, and one to Peggy's Cove. "Now that," he was quoted as saying later at a debriefing on emergency preparedness, "you can't put in a plan."

ASIDE FROM HAVING HUGE DISASTERS such as the Swissair to deal with, the fire services continued to be faced with large fires within the city throughout the 1990s. Some of those fires even raised suspicion. The Starr Manufacturing fire in Dartmouth on June 12, 1998, was one such incident when an early morning blaze gutted one-third of the historic building in downtown Dartmouth. A day after the fire, investigators told the *Daily News* that arson was suspected. They had also discovered a fire at the nearby Greenvale School. The two sites were vacant and had been entangled in a tug-of-war as both Sobeys and Loblaws hinted at interest in buying the properties. Some local residents and businesspeople had voiced publicly their concerns that the buildings' historic values should be maintained.

Firefighters were called to the Prince Albert Road factory at 5:30 A.M. after fire started on an outside corner of the building. In the end, an exterior wall and part of the roof were destroyed, and it took more than two hours to get the fire under control. Halifax Regional Fire Inspector Tom Silver later told the *Daily News* that there was no evidence that an accelerant, such as gasoline, was used. He estimated the damage at $150,000.

The factory, which at the turn of the century manufactured eleven million skates for a generation of North American and European children, as well as the King of Spain, closed in June 1996.

ON DECEMBER 31, 1998, a tragic fire at Sunrise Manor on Gottingen Street in Halifax left two women dead and twenty-eight others in hospital suffering from smoke inhalation. Careless smoking sparked the fire in the senior citizens' residence, rapidly filling the building with smoke. The ten-storey, 166-unit building was reportedly not equipped with sprinklers.

The following year, a fire in Halifax's southend neighbourhood destroyed almost half a city block. According to regional dispatchers, firefighters were called to 5240 Smith Street at 12:09 A.M. on December 12, 1999 after a tenant heard "pops and crackles" in the residence. Three fire trucks, from University Avenue, West Street and Lady Hammond Road, arrived at about 12:15 A.M. with twelve firefighters. Upon arrival, the on-scene captain immediately called for two more trucks with eight firefighters.

The crews tackled the blaze and within about an hour, it appeared as if they had suppressed the fire. It was only after they were checking to see if it had spread to other buildings that they realized how big a problem they had. The house at 5240 Smith Street was one of a row of six historic Victorian homes. The row houses, built for tannery workers in around 1870, were renovated about thirty years ago, creating one of the prettiest streets in the city. The common roof covering the row provided a path for flames to creep from house to house in both directions, and a lack of firestops in the old balloon framing, in which studs in the homes ran from the basement to the attic, let fire spread quickly between floors. Today, the floor supports in a home act as a barrier between storeys, hampering fire from spreading between levels.

Firefighters had to tear down one of the row houses to help stop the fire's spread.

Fire inspectors estimated that the fire caused upwards of two million dollars in damages, including the buildings' contents, and left thirty-two people homeless. Three firefighters were taken to hospital: one fell about three metres from a ladder and was treated for a sore shoulder and released, and two others were knocked down by an apparent explosion while on a roof. They were released without injuries.

Fire investigators believed electrical problem involving computer equipment caused the fire. After sifting through the charred remains, they

Firefighters battle a house fire in Halifax, 1998

told the *Chronicle Herald* they thought the fire started at or near an electrical outlet in a den on the main floor.

After the fire, some area residents questioned whether enough firefighters had responded to the blaze, speculation that renewed a debate about adequate staffing. Since amalgamation, the fire service had lost about sixty career firefighters, mostly to retirement. The Halifax Regional Professional Firefighters Association wanted them all replaced, saying that large areas were understaffed, especially during major fires when firefighters had to be drawn from Bedford, Dartmouth and other communities. Earlier in 1999, the union handed Halifax Regional Council a report about its staffing concerns which focused particularly on the fatal New Year's Eve fire at Sunrise Manor. During that fire large parts of the municipality were seriously understaffed, the report said. Chief director Michael Eddy told the *Chronicle Herald* at the time that the department was adequately staffed for day-to-day necessities, but conceded that major fires stretched resources too thin.

NOT ALL FIRES, of course, are accidental. The act of deliberately setting a fire—arson—has been a crime firefighters have had to battle since the earliest days. Whether motivated by insurance payouts, vengeance, or adolescent mischief, arsonists continue to pose a problem for firefighters everywhere.

In 2000, four high-school students were charged with firebombing Prince Andrew High School in Dartmouth the week before Christmas. The fire was started by two or three Molotov cocktails, according to regional fire officials. The homemade devices were believed to have been thrown through a window. The December 18 blaze was quickly doused by firefighters and the school's sprinkler system, but smoke and water damage was heavy in the library and cafeteria. Fire officials suggested that the repair bill could be between fifty and one hundred thousand dollars.

The following year on November 27, firefighters were called to the two-hundred block of Wyse Road in Dartmouth where flames had engulfed a warehouse, sign store, dozens of cars at a nearby auto dealership, and the Old Mill Tavern. The spectacular blaze, for which arson was blamed, took firefighters several hours to bring under control.

Less than a month later on December 18, firefighters in Halifax scrambled to contain seven deliberately set fires in a five-block radius south of Spring Garden Road. Fortunately, none of the fires got out of control. Area residents dubbed the incident the "night of flames."

In 2001, recycling plants proved to firefighters their potential to be fire traps. On June 11, a call came in from 5 Brown Avenue in Burnside Industrial Park where bundles of cardboard stored at the recycling plant had ignited, forcing the evacuation of several nearby businesses. Flaming debris scattered throughout the park, causing firefighters to scramble to contain small spot fires. The fire burned for several hours before it got under control. No injuries were reported, but an excavator operator at the fire suffered a heart attack and later died in hospital. In June of the following year, a serious fire broke out at a recycling depot in North Preston. The fire, which started in a five-metre-high mound of construction and demolition material, burned for forty-eight hours, threatening nearby woods. Thirty-five firefighters from six stations in the area battled the fire as a helicopter dropped water from a nearby lake on the flames. More than forty million litres of water are reported to have been dumped by firefighters on the fire before it was eventually extinguished.

Burnside Recycling Fire, 2001

Fires and Disasters

Responding to the September 11, 2001 terrorist attacks

CAPTAIN LYE AND FIREFIGHTER PAUL EDWARDS of the Halifax Regional Fire and Emergency Services visited Ground Zero in New York City a couple of months after the terrorist attacks to deliver a $27,000 cheque. Local 268 of the International Association of Firefighters also sent a $33,600 cheque to their fellow union members in New York. More than three hundred members were killed in the line of duty in the wake of the terrorist attacks in New York City.

In the hours after the attacks in New York City and Washington, D.C., Halifax received more than seven thousand stranded air travellers coping with fear, anxiety, and fatigue. The international passengers came from forty planes that had been diverted to the city after the Federal Aviation Administration halted all flight operations at US airports.

The fire department was called in to play a non-traditional role. There were no fires, hazardous spills, or motor vehicle accidents to attend to, but a large number of people needed to be organized and provided with care. The fire service stepped in to help other government agencies and aid organizations such as the Red Cross. Firefighters arrived at the airport to get stranded passengers onto Metro Transit buses which had been provided by the city to transport them to temporary shelters. On the bus, the sight of uniformed firefighters helped to reassure the anxious travellers. The firefighters used the travelling time on the bus to gather names and identifying information to ensure that everyone was not only accounted for, but had a place to sleep and a meal to eat. It would be several days before all the stranded passengers were able to continue on their journeys.

THE YEAR 2003 proved to be another particularly eventful year for Halifax firefighters. It started on February 26, 2003, when firefighters from across Metro struggled for hours in freezing temperatures to contain a massive fire that swept through four buildings on Gottingen Street. Fearing that the fire would spread through the dense block of mostly wooden and stone houses, firefighters ordered an evacuation of the area surrounding the fire.

The blaze was started by a roofing company's propane torch during repairs to the old wood and stone buildings. The torch heated up a nearby wall space, and the smoldering spread throughout the building's walls, floors, and ceilings for forty-five minutes before any flames were visible. Forty-five firefighters worked the scene, with seven engines, three aerial trucks, a mobile command post, and one tactical unit.

Firefighters not only had to contend with frigid temperatures but also no fire walls, which allowed the flames to spread rapidly in the old buildings. Frozen fire gear also posed a serious problem. "They have nearly a mile of frozen hose down there they have to thaw out," fire department spokesman Mike LeRue told the *Daily News*.

Firefighters doused the fire with water and foam, which helps water penetrate into wood, for hours. An estimated four-and-a-half million litres of water was dumped on the fire. By the time excavators moved in to knock down the fire-ravaged buildings, a dense, sooty ice covered overhead power lines and the street below, posing a significant risk to firefighters and their equipment.

The fire left at least twenty people homeless and destroyed a hair salon, a tattoo and piercing parlour, a used-clothing store, several apartments, and a home.

Gottingen Street fire, 2003

LATER THAT YEAR on the morning of August 7, 2003, a powerful explosion at the Halifax Grain Elevator Limited in southend Halifax forced the evacuation of up to four hundred residents. Streets were shut down as bitter-smelling smoke filled the air. Fortunately, no one was injured in the 11:45 A.M. blast and residents were allowed to return home that evening. Jasen Fisher, a resident of the area, described the scene to the *Chronicle Herald*: "The house literally shook. My wife and I initially thought that a transport truck had run into the side of the house—that's how heavy it was."

Witnesses reported seeing a fireball shoot into the sky and hearing a boom that shook homes and rattled windows. An outer wall was blown out of the six-storey grain elevator, and sheet metal was ripped away from the

Grain Elevator Explosion, 2003

building. Two dozen firefighters spent the day struggling to extinguish the blaze, while smoke billowed from the building where wood pellets, grain, and milling wheat were stored. Firefighters soon got the fire under control, but it smoldered for several days.

The explosion was caused by the ignition of grain dust, which can be extremely flammable. "Moist conditions make grain dust even more dangerous. Grain dust is kind of an odd creature. It's an organic compound so when the humidity is high, the ability of it to spontaneously ignite increases, the same as wet hay," Bruce Burrell, the acting chief director for the Halifax Regional Fire Service, reported at the time. A fire investigation report released in December 2003 determined that the explosion was likely caused by poor maintenance and friction from worn bearings. It was believed that flammable dust somehow caught fire and caused the blast.

In September 2004, all seven silos at the Halifax grain elevators storing wood pellets were deemed a fire threat and ordered to be emptied. Workers cleared the pellets out of the silos under the watch of the fire service. The concern was that they were heating up in the silos and could cause another explosion.

AT 12:10 A.M. ON SEPTEMBER 29, 2003, Hurricane Juan made landfall in Nova Scotia somewhere between Shad Bay and Prospect. High winds, with gusts of up to 176 kilometres per hour, toppled huge trees and power poles and ripped roofs off of buildings. Storm surges pushed boats and wharfs up onto the shore. Hundreds of thousands of Maritimers in Nova Scotia and

Prince Edward Island lost power in the storm and remained without it for days—and for some, weeks—afterwards.

Juan claimed the lives of eight people. Two were killed when trees fell on their motor vehicle, two fishermen died near Anticosti Island, and three died in a house fire that is believed to have been started by candles used during the power outage. The other death occurred during relief work in the weeks following the storm. The last time Halifax was hit by such a severe hurricane was on August 22, 1893, when a storm made landfall in St. Margaret's Bay at about 3 A.M. That storm, known then as the "second Great August Gale," claimed twenty-five lives.

Halifax Regional Municipality declared a state of emergency and ordered people living close to the shore in low-lying areas to seek alternative shelter before the storm made landfall. Emergency shelters were set up at local fire halls for people forced out of their homes. Efforts were all co-ordinated through the EMO command centre in Dartmouth, not far from the Nova Scotia Hospital. In a room filled with overhead projectors, phones, and writing boards, emergency workers scrambled to co-ordinate rescue efforts and dispatch information.

When the roof blew off a building on Windmill Road, fire engines raced to the scene to provide residents with their trucks for use as a temporary shelter, while EMO organized Metro Transit buses to transport them to a more secure shelter. "It was a tremendous experience of working together," said Barry Manuel.

Hurricane Juan brings havoc, 2003

Because of the lack of power in the days following the hurricane, water from drilled wells was unavailable. The fire service announced that drinking water was available at twenty fire stations throughout the municipality. In some rural fire stations, the doors were opened in the days after the storm, offering hospitality to those without power. At the station in Oyster Pond, the ladies' auxiliary offered food, tea, coffee, and company at the hall.

On the night of September 29, firefighters across the region left their own families and put aside concerns for their own safety to help others. In Bedford, the pagers of all volunteers started ringing. They were asked to report to their stations for emergency standby. They weren't there long when they were told they had to start evacuating residents on Union Street and Shore Drive, Lieutenant Robert Andrews recalled. The rain and wind had started, water levels were quickly rising, and trees were beginning to fall on power lines. During the height of the storm, the crew was on Dartmouth Road when a high-voltage wire caught on fire, and the fire truck driver had difficulty getting traffic to stop nearby, Andrews recalled.

In Musquodoboit Harbour, the first call came in from the chief's wife; a transformer had blown right outside their house. The chief, along with some firefighters, went to investigate. All night the station answered calls to block highways where power lines were down and to insure one part of highway remained open for emergency vehicles. The day after the storm, volunteers were out with chainsaws helping neighbours clear trees and debris. The station provided a generator and running water for the community, along with freshly-brewed coffee, and meals for those in need. For the next several days, firefighters and their families delivered water to seniors and attended medical assist calls.

The fire service was flooded with calls in the hours and days after the hurricane. From September 28 until October 12, the day Nova Scotia Power restored service to all customers, the HRM fire service received 1,495 calls. During that same period in 2002, it received 636. While many of the calls were attributed to alarm malfunctions, others were blamed on kids lighting piles of branches or the misuse of candles and oil lamps.

Tragedy struck a westend Halifax family in the early morning hours of September 30. A mother and two children, ages two and four, died in a fire at their home as the children's helpless father and neighbours watched in horror. Neighbours spotted the fire and managed to wake father Bill Amro from his sleep. He got his five-year-old daughter, Anna, out through a bedroom window and onto an overhang above the front door where a neighbour lifted her to safety. The father tried to go back for the rest of his family, neighbours told the *Daily News*, but the fire and smoke were so intense he had to jump to safety.

It took fire engines only five minutes to reach the public housing development, but it was already too late—the home was engulfed in flames. Firefighters tried in vain to revive the first child pulled from the blaze, but it was evident the other child and the mother were already dead. The Amros' fire detector didn't go off because it was wired into the electrical system, as the building code required, rendering it useless after the storm knocked the power out. Fire investigators later determined that the fire was accidental, but they were unable to conclude whether it was started by cigarettes or candles. The fatal fire was the second blaze that day for city firefighters; a candle left beside a bed seriously damaged an apartment on Quinn Street in the west end, but no one was injured.

"REGIONAL DISPATCH TO ZONE 4, we have a report of a plane crash just off the runway at Halifax International Airport. There are multiple fires burning. The aircraft is a 747, time out, 4:01(A.M.)."

These were the first words most firefighters in the Waverley to Beaverbank area heard in the early morning of October 14, 2004. HRM firefighters, along with airport fire trucks, were on-scene within minutes and battling the intense fire. District Chief Blois Currie was the first member of the regional fire service to arrive. He advised dispatch he was on-scene and confirmed that there were multiple fires burning and heavy smoke. While attempting to take off, the Boeing 747 cargo plane had cut through the nearby forest before crashing and flying apart. The crash killed all seven crew members onboard the plane which was loaded with tractors, computer gear, and seafood. When firefighters first arrived, they feared that they would find a passenger plane. "It was forty-five minutes to an hour before we found out that it wasn't," Chief Currie told the *Chronicle Herald*.

Chief Currie told the newspaper that he didn't have much time to absorb the situation before him: "What was going through my mind was manpower, water supply, where we were going to set up and get our personnel accounted for, and probably about four hundred other things. Through training, whether it's a house fire or a plane crash, we set up all the same way, and we do it over and over and over again."

When Waverley Deputy Chief Steve Comeau arrived on the scene, he volunteered to do a perimeter check. "During the initial walk I encountered heavy fire and heavy fuel flames within the debris and the trees were burning but not taking hold. I also looked in the trees and on the ground for victims and found none," he described. "I smelt fuel and noticed small fires within the clearing the aircraft made when it touched down. I crossed the Old Guysborough Road and noticed airplane parts on the road; the power wires

Plane crashes at Halifax International Airport, 2004

along with a power pole were broken. I was able to tour the area where the aircraft struck the runway and the landing systems and notice tail debris and cargo contents in the trees beyond the runway."

Firefighters quickly found a water supply from a hydrant not far from the plane's path, and a five-inch feed line was run to airport and HRM fire engines close to the fire. Firefighters, hampered by poor visibility and smaller fires fuelled by two hundred thousand litres of spilled jet fuel, worked for almost four hours getting the fires under control and dousing hot spots. Despite the charred wreckage, the scattered razor-sharp pieces of aircraft parts, and hazardous, uprooted trees, no injuries were reported.

Nearly one hundred firefighters—ninety per cent of them volunteers—responded to the crash, said Currie. Along with members from stations in Waverley, Fall River, and Wellington, firefighters also came from the nearby Elmsdale and Enfield departments. Other departments, such as Cole Harbour, remained on standby.

WHATEVER THE TYPE OF CALL, the fire service is kept busy. In 2004, the eighteen stations in the region's urban core responded to 9,865 calls which ranged from structure fires to medical emergencies to activated alarms. Meanwhile, firefighters in forty-three rural stations responded to 3,665 calls. Station No. 2 on Halifax's University Avenue, next door to the IWK Health Centre, is the busiest station in HRM, having responded to 1,235 emergency calls in 2004. Among the rural departments, Prospect Road and District was the busiest, with two hundred and ninety-nine calls.

Firefighters' commitment does not end when the emergency is under control and trucks are back in the station. Helping their communities through fundraising is another important part of many firefighters' lives. One good example is the Nova Scotia Firefighters' Burn Treatment Society, founded in 1983 by eight firefighters in the province. The charitable organization's aim is to raise money for the province's two burn-treatment centres at the QEII Health Sciences Centre and the IWK Health Centre.

Almost a decade after amalgamation and the creation of the Halifax Regional Fire and Emergency, the service is moving into a new phase, one that is clearly focused on the future. With the municipality in the process of developing a long-range, region-wide plan that will outline where, when, and how future growth and development should take place within HRM, the fire service is working with municipal planners to ensure new growth areas in the city have adequate levels of fire protection and service.

But despite all the changes within the region's fire service, some things stay the same; notably, the dedication required to do the challenging job. Firefighters are among the most trusted and most well-liked professionals around, and it's not hard to see why. Barry Manuel, HRM's Emergency Measures Organization co-ordinator, puts it plainly: "I've never met a group of people with more compassion."

Firefighter Wayne Ledson gives oxygen to a dog at a fire on Old Sambro Road, c. 1986

Chapter 6

Then and Now

In the photo to the left, the fireman's embroidered lapel, tie, and cummerbund are much more elaborate than the relatively plain designs seen in other photos of firefighters taken around the same time. This might indicate that the man in the photo held a higher position than that of the general rank-and-file. He clutches an ornamental horn, which rests on the table beside him. To be heard above the chaos that marked fires in the 1880s, officers shouted directions to their firefighters through a silver speaking-trumpet. In addition to their practical role, the trumpets were also used as ceremonial trophies. Engraved trumpets were given as awards at firemen's tournaments. Today's fire service wears trumpet "collar dawgs" to denote the rank of officers.

Fireman's uniform, Halifax, c. 1880

A modern firefighter, 2005

NADYA PARE is one of the seven female career firefighters in Halifax Regional Fire and Emergency. Hired in May 1994, she is posted at the Knightsridge Drive station in Halifax. After more than a decade in the fire service, Pare is often called upon to speak publicly about what it's like to be a female firefighter. In the beginning, she says, it was challenging. Working in a traditionally male-dominated environment meant she initially had to share a bathroom with her male colleagues and wear work clothes that were designed to fit men. Through education and training, things are much better today, she says. The skin magazines and posters are gone from the fire hall walls, and women are more accepted and respected by the male firefighters. With even more women entering the region's fire halls, the department has had to adapt and is moving toward building separate sleeping quarters. Currently, paid female firefighters working on twenty-four-hour shifts have no choice but to stay in the same sleeping area as the men.

Regardless of gender, today, each firefighter is equipped with a set of bunker gear that consists of three-layered jacket and pants, much more comfortable and breathable than the long rubber coats worn in the early 1990s. The outer shell is made of a fire-resistant blend of a material called Nomex Kevlar, and the inner liner is also fire-resistant but also provides thermal protection. In addition, there is a water-resistant vapour barrier which allows for breathability. The gear also includes a flash hood, gloves, bunker boots, a helmet, and ear defenders. External gear includes a hand light, a spanner wrench for coupling hoses, door and sprinkler stops, rope, screwdrivers, knives, and window punches. On their backs, firefighters wear a bottle which provides air for a self-contained breathing apparatus. The bottles can provide anywhere from twenty minutes to one hour of air. The breathing apparatus, which makes the job much safer, has been part of the firefighters' equipment since at least the mid-1980s.

Engine House on Halifax's Spring Garden Road, 1886

THE FIREMEN POSE with a decorated hose reel cart and Smith hand engine at the Spring Garden Road fire hall during the 1886 Grand Firemen's Tournament. The station, located at the corner of Spring Garden Road and Brunswick Street, closed in 1920. It was recommended to close in the mid-1890s, but in 1916 it was still operating as a hose company with two paid staff, nine men on call, a wagon and horse, six hundred feet of hose, and one three-gallon fire extinguisher.

Highfield Park station, 2004

LOCATED AT 45 HIGHFIELD PARK Drive in Dartmouth, the Highfield Park Fire Station is Halifax Regional Fire and Emergency's newest.

The opening ceremony of the new fifteen-hundred-square metre structure was held on October 13, 2004. Mayor Peter Kelly and Councillor Jim Smith cut the ribbon. Smith and Chief Director Mike Eddy also unveiled the station's dedication plaque. Built for three million dollars, the station has three drive-thru bays, which can each accommodate two trucks.

Brunswick Street fire, 1934

A SUSPICIOUS EXPLOSION took place at 214 Brunswick Street on October 11, 1934, killing seven people. In the photo to the left, firemen examine the rubble left behind.

Halifax's enviably clean fire record in the mid-1800s crumbled during a sixty-year period between 1890 and 1950. On October 1, 1891, the "Great Waterfront Fire" took place. Described as the "most disastrous in thirty years," it was fuelled by huge stores of petroleum oil. Shops, offices, warehouses, and wharves went up in smoke. Several other major fires occurred during the sixty-year period, including the Pier 21 blaze on March 5, 1944. No lives were lost but damages were reported at two hundred thousand dollars.

A LATE-NIGHT FIRE gutted a three-storey apartment building in Halifax's Rockingham area in March 1994. The fire broke out at about 10:45 P.M. at 6 Langbrae Drive. More than a dozen people had to be rescued, the *Chronicle Herald* reported. Seven fire stations and twenty-five firefighters were on the scene until close to 3:30 A.M. No major injuries were reported, but one firefighter was taken to hospital with a minor leg injury and later released.

Rockingham apartment building fire, 1994

Firefighters enjoy Christmas dinner L to R: John Crosby, Edgar Kinghorne, Berkley Smith, Cadt Hartley Hushard, Ray Wambolt, Fred Hencher, Fred Findlay, Matt Lynch and Captain Nelson Cormier enjoy a Christmas Day dinner at the old Oxford Street Fire Station in Halifax.

POKER, CRIBBAGE, AND POOL were once common recreational activities in fire halls. Today, they have been replaced by televisions and exercise equipment. Firefighters are encouraged to spend some of their downtime keeping fit. When Fred MacGillivray became fire chief in 1945, he made some changes to how firefighters spent their time when they weren't answering calls. In a retirement tribute, the former chief was quoted by the *Mail Star* on January 2, 1962, as saying that he had "determined that it was time the fire staff spent their time constructively on classes and maintenance. The familiar cartoon of firemen sitting around a poker table waiting for the bell has no place in Halifax."

L to R: Mark Boone, Chris Sweet, Captain Joe Ryan, and Nick Morash enjoying a communal meal at the University Avenue Station in Halifax

WITH TODAY'S FIREFIGHTERS working twenty-four-hour shifts with the same crew, week after week, being able to maintain harmony with fellow employees is an essential part of the job. "[The] fire [service] is trained to work together as a team," says Barry Manuel of HRM's Emergency Measures Organization. "They train as a family. They work as a family. They fight as a family."

Bill Mosher, deputy chief of rural operations, joined the Halifax department in October, 1978. He was a firefighter for ten years before moving into management. "Going to the calls was the best part," he said. Like a lot of firefighters, Mosher, who hung around fire stations as a kid with his firefighting father Henry Mosher, was known as a prankster. "We had a lot of fun." The pranks he and others were involved in ranged from pouring buckets of water over unsuspecting heads, flouring a bed (covering the sheets with flour so that when someone laid down they would get covered head to toe in white flour), and faking a fist fight between two firefighters, so that when the district chief came over to mediate, he'd find out it was all a joke.

A ladder demonstration at a Halifax fair, c. 1940

FIREMEN DEMONSTRATE their agility on raised ladders to spectators. In the early 1900s, firefighters would commonly create ladder arches over streets to celebrate special events. One was erected on the corner of Cunard and Gottingen Streets to welcome the Prince of Wales to Halifax on August 19, 1919.

250th anniversary celebrations, 2004

As part of its anniversary celebrations, the Halifax Regional Fire and Emergency Service partnered with Grades 3–6 of Colonel John Stuart Elementary School in Cole Harbour. The goal was for the students to develop an artistic design representing the fire service. Grade 6 student Anslie Cunningham produced the winning design.

Two hundred and fifty years after the first official fire service was established in Halifax, fire protection in not only the city, but also the dozens of surrounding communities, has come a long way. Today the fire service operates out of 62 stations across the region (as well as its central headquarters in Dartmouth), making it a truly regional service.

Firefighters' responsibilities have grown and become more complex over the years. A highly-trained profession, firefighting is only one element of their duties: they must deal with serious motor vehicle accidents, respond to hazardous chemical spills, and work with other regions across the country on chemical, biological, radioactive, and nuclear response strategies.

250th anniversary celebrations

AT THE 250-year anniversary celebrations held at the Grand Parade in Halifax on May 31, 2004, a new flag was unveiled and the lieutenant governor unveiled a statue of a firefighter made from a tree which fell during Hurricane Juan. The statue was carved by District Chief Paul Hopkins.

The fire service's history has been preserved thanks in large part to a group of former paid and volunteer firefighters in the region who have collected pieces of fire service equipment and vehicles. Motivated to preserve the history of local fire services, they established the Regional Firefighters Interpretation Centre. The artifacts they have collected include antique fire engines, old fire extinguishers, breathing apparatus, radio-communication equipment, fire service-related photos, and paintings. In 2004, they set up a display in the Bedford Place Mall in Bedford. The former firefighters were excited to be able finally to put some of the items, some of which had spent years in storage at scattered sites in Lower Sackville, Beaverbank, Halifax and Eastern Passage, on display. The centre's goal is to find a permanent property, preferably a vacant fire station, where people could go to learn about the region's firefighting history. The province has a firefighting museum in Yarmouth.

THE HALIFAX REGIONAL FIRE and Emergency's fire engines today are anywhere from thirty to thirty-two feet long, ten feet high and eight feet wide. They cost about $450,000 and carry 750 gallons of water, 25 gallons of foam, and are equipped with a 1,750 gallons-per-minute pump. On the highway, they can reach a speed of sixty miles per hour but for the most part will refrain from going that speed.

AT ONE TIME, a fire pole could be found in every engine house. At the sound of an alarm, the pole allowed firemen to move quickly from their sleeping quarters above to the fire apparatus below. Today, the poles are not as common, but can still be found in some stations including Station No. 2 on University Avenue in Halifax.

Modern apparatus

Racing to a call,
January 17, 1953
TOP TO BOTTOM: Jackie Joseph, Doug Keeping, and Ken Walker

Practicing with a life net in front of the Bedford Row station in Halifax.

THE JOB OF A FIREFIGHTER is extremely demanding physically, requiring high levels of cardiopulmonary endurance, muscular strength and endurance. Training has been a part of every fire department's routine since the very early days. It has obviously developed over time and become much more organized and formalized.

ICE RESCUE is an important part of a firefighter's ongoing training. In January 1987, Dartmouth firefighters were called to use their skills after a twenty-two-year-old man got trapped under the ice in Lake Banook. Divers from the city's fire and police departments, the RCMP, and military took

Firefighters practice an ice rescue, 1987

part in the two-day search for the diver. Division Chief Joseph Vidito of the Dartmouth Fire Department, said it appeared the man became lost and swam away from the open waters of the lake's southern end. The body was found about a thousand yards from the opening, twenty feet from the shore near the Micmac Aquatic Club. "It's a tragic thing what's happened here," Vidito told a local reporter.

Prior to 1996, each department took care of its own training. After amalgamation, training came under one division of the Halifax Regional Fire and Emergency. Today, David Burnet, the divisional chief of training for the fire service, has ten people in his division, including seven training officers. Firefighters undergo annual training to keep fit, learn new techniques, and maintain their existing skills.

He explains that all the new career firefighting recruits must have a Level 1 National Fire Protection Association (NFPA) training certificate. Obtaining that certificate costs upwards of nine thousand dollars. After a firefighter has a NFPA Level 1 training certificate, they will go through a Candidate Physical Ability Test (CPAT). "It tells us you're fit to do the job," says Burnet. Part of the test includes a stair climb designed to simulate the critical task of climbing stairs while carrying an eleven-kilogram hose pack in full protective clothing. Another test simulates rescuing a person from a fire scene. The firefighter has to grab a seventy-five-kilogram mannequin and drag it for more than twenty metres.

Last Alarm

The following members of what is now Halifax Regional Fire and Emergency have given their lives while serving their community. There are likely more firefighters who died in the line of duty who are not listed.

Lieutenant **Edward Fredericks** of Halifax died on April 14, 1878
Firefighter **Rufus Keating** of Halifax died in January 1894
Lieutenant **William Lewin** of Halifax died in March 1898
Firefighter **James O'Regan** of Dartmouth died in February 1901
Hoseman **Michael Sullivan** of Halifax died on March 21, 1903
Chief **Edward Condon** of Halifax died on December 6, 1917
Deputy Chief **William Brunt** of Halifax died on December 6, 1917
Captain **William Broderick** of Halifax died on December 6, 1917
Capt. **Michael Maltus** of Halifax died on December 6, 1917
Hoseman **John Spruin** of Halifax died on December 6, 1917
Hoseman **Walter Hennessey** of Halifax died on December 6, 1917
Hoseman **Frank Leahy** of Halifax died on December 6, 1917
Hoseman **Frank Killeen** of Halifax died on December 6, 1917
Hoseman **John Duggan** of Halifax died on December 6, 1917
Hoseman **William Gorman** of Halifax died on February 14, 1926
Hoseman **William Cormier** of Halifax died on April 15, 1930
Hoseman **William Knapman** of Halifax died on January 25, 1939
Firefighter **Will Boston** of the Fairview Fire Department died in March 1954
Captain **Earl Fox** of Halifax died on January 15, 1956
Captain **Richard Keily** of Halifax died on December 12, 1960
Lieutenant **William (Billy) Carter** of Halifax died on December 3, 1973
Firefighter **Allen MacFarlane** of Halifax died on January 20, 1980
Firefighter **George Joseph Branch** of Halifax died on November 24, 1980
Firefighter **Raymond Kline** of Halifax died on December 23, 1981
Firefighter **William Ernst** died on January 18, 1990
Firefighter **Ronald MacDonald** of the Lake Echo Fire Volunteer Department died on November 2, 1997

Image Sources

Dartmouth Heritage Museum: 21, 91

Halifax Regional Fire and Emergency Service: cover, map, introduction, 3, 4, 7, 9, 15, 16, 17, 20, 24, 27, 29, 31, 37, 38, 42, 43, 46, 49, 50, 51, 53, 54, 55, 57, 58, 62, 63, 64, 65, 67, 68, 69, 70, 72, 73, 75, 76, 77, 78, 79, 80, 81, 82, 83, 85, 89, 90, 92, 93, 94, 95, 96, 98, 99, 100, 101, 104 (Doug Leahy), 106, 107, 109, 111, 112, 113, 114, 121, 123 (Bruce MacDonald), 124, 125, 126, 130, 131, 134, 136, 137, 138, 139, 140, 141, 142, 143, 144, 145

Maritime Command Museum: 25, 26

Museum of Natural History: 32 (75.70.3)

Nova Scotia Museum: 12

Nova Scotia Archives and Records Management: 6 (Notman 16804), 8 (1979-147.24), 11 (N-877), 28, 30, 35, 40 (John Rogers N-9453), 47 (Yarmouth County Historical Society N-639), 48, 88, 135 (Tom Connors 648),

Private Collections: Barb Sawatsky, 59; Wilma Brimicombe, 60; Gary Castle, 67; Murray Elliott, 71; George Crook, 115

Chronicle-Herald: 118 (Tim Krochak); 127 (Darren Pittman)

Dalhousie University: 97

Bibliography

Africville Genealogical Society. *The Spirit of Africville*. Halifax: Formac Publishing Company Ltd.,1992

Baird, Donal. *The Story of Firefighting in Canada*. Erin, Ont.: The Boston Mills Press, 1989.

Bassett, John M. *Samuel Cunard*. Don Mills, Ont.: Fitzhenry & Whiteside Ltd., 1976.

Burgess-Wise, David. *Fire Engines and Firefighting*. Norwalk, Connecticut: Longmeadow Press, 1977.

Chapman, Harry. *Along the Cole Harbour Road: A Journey Through 1765–2003*. Dartmouth: Cole Harbour Heritage Society, 2003.

———. *In the Wake of the Alderney: Dartmouth, Nova Scotia, 1750–2000*. Dartmouth: Dartmouth Historical Association, 2000.

Clairmont, Donald. *Africville Relocation Report*. Halifax: Dalhousie University, Institute of Public Affairs, 1971.

Edwards, Don and Devonna. *The Little Dutch Village: Historic Halifax West: Armdale, Fairview*. Halifax: Nimbus Publishing, 2003.

Erickson, Paul. *Historic North End Halifax*. Halifax: Nimbus Publishing, 2004.

Fingard, Judith, Janet Guildford and David Sutherland. *Halifax: The First 250 Years*. Halifax: Formac Publishing Company Ltd., 1999.

Kimber, Stephen. *Flight 111: The Tragedy of the Swissair Crash*. Toronto: Seal Books, 1999.

Kitz, Janet. *Shattered City: The Halifax Explosion and the Road to Recovery*. Halifax: Nimbus Publishing, 1989.

Ledger, Don. *Swissair Down: A Pilot's View of the Crash at Peggys Cove*. Halifax: Nimbus Publishing, 2000.

Marshall, Dianne. *Georges Island: The Keep of Halifax Harbour*. Halifax: Nimbus Publishing, 2003.

Martin, John Patrick. *The Story of Dartmouth*. Dartmouth: Privately Printed by the Author, 1957.

McCreath, Peter L. *The Life and Times of Alexander Keith: Nova Scotia's Brewmaster*. Tantallon, N.S.: Four East Publications, 2001.

Nova Scotia Archives and Records Management. *Halifax and its People 1749–1999.* Halifax: Nimbus Publishing, 1999.

Pacey, Elizabeth. *Miracle on Brunswick Street: The Story of St. George's Round Church and the Little Dutch Church.* Halifax: Nimbus Publishing, 2003.

Pacey, Elizabeth and Alvin Comiter. *Historic Halifax.* Willowdale, Ontario: Hounslow Press, 1988.

Parker, Mike. *The Smoke-Eaters: A History of Firefighting in Nova Scotia 1750–1950.* Halifax: Nimbus Publishing, 2002.

———. *Historic Dartmouth.* Halifax: Nimbus Publishing, 1998.

Payzant, Joan. *Like a Weaver's Shuttle: A History of the Halifax–Dartmouth Ferries.* Halifax: Nimbus Publishing, 1979.

Raddall, Thomas. *Halifax: Warden of the North.* Halifax: Nimbus Publishing, 1993.

Saunders, Charles. *Black and Bluenose: The Contemporary History of a Community.* Lawrencetown Beach: Pottersfield Press, 1999.

Stephens, David. *It Happened at Moose River.* Hantsport: Lancelot Press, 1974.

Withrow, Alfreda. *One City Many Communities.* Halifax: Nimbus Publishing, 1999.

———. *St. Margaret's Bay: An Historical Album, Peggy's Cove–Hubbards.* Halifax: Nimbus Publishing, 1997.

Wright, H. Millard. *The Other Halifax Explosion: Bedford Magazine July 18–20, 1945.* Halifax, 2001.

Booklets, Papers, Magazines, Video

Annual Reports of the Halifax Fire Department and Halifax Regional Fire and Emergency Services 1900–2004.

Annual Reports from the Dartmouth Fire Department.

Beaverbank Kinsac Volunteer Fire Department. *Twenty-five years of Community Service 1965–1990.*

Connelly, Pearl. *The Bicentennial of the Halifax Fire Department 1768–1968.*

De Long, Jodi. "In the Line of Fire." *Saltscapes* (March/April 2005): 43–48.

FeedLine. A newsletter dedicated to members of the Halifax Regional Fire and Emergency (various issues).

Halifax Fire Department. *An Historical Celebration: 225 Years of Firefighting in Halifax, 1768–1993.*

Lakeside Volunteer Fire Department. *Fiftieth Anniversary 1948–1998.*

Jost, Philip. *Halifax Regional Fire and Emergency Service: Rural District Fire Service Management Review and Recommendations.* Executive Master of Business Administration Program at Saint Mary's University, 2001.

Hayes, Charles J. A. "Swissair Flight 111 Crash." *Canadian Journal of Emergency Management* (May/June 1999): 20–24.

Kehoe, Pete. "Swissair Aftermath." *Firefighting in Canada* (May 1999): 4–11.

Sackville and District Fire Department. Annual report, 1984.

Newspapers

The following newspapers, circa 1840–2005, provided valuable research material. Some of the clippings used are contained in archival and museum holdings, while others were received from private collections. Exact sources have been given in some parts of the book's text.

Halifax Morning Sun, Daily Echo, Acadian Recorder, Morning Post, Halifax Evening Express, Halifax Reporter, Halifax Mail, Chronicle Herald, Mail Star, Halifax Daily Star, Daily News, Dartmouth Free Press, Globe and Mail.